Scientific Enigmas

Scientific Enigmas

ROBERT B. DOWNS

1987
LIBRARIES UNLIMITED, INC.
Littleton, Colorado

LIBRARIES UNLIMITED, INC.
P.O. Box 263
Littleton, CO 80160-0263

Library of Congress Cataloging-in-Publication Data

Downs, Robert Bingham, 1903-
Scientific enigmas / Robert B. Downs.
xiii, 153 p. 17x25 cm.
Includes bibliographies and index.
ISBN 0-87287-617-9
1. Science--Miscellanea. I. Title.
Q173.D68 1987 87-26870
001.9'4--dc19 CIP

Libraries Unlimited books are bound with Type II nonwoven material that meets and exceeds National Association of State Textbook Administrators' Type II nonwoven material specifications Class A through E.

For Jane Bliss Wilson Downs,
who originally inspired the present work.

Contents

Preface

The original impulse for the present work developed out of research into the history of science, represented by my *Landmarks in Science* (Libraries Unlimited, 1982) and perhaps by a naturally skeptical turn of mind. The true scientific spirit is epitomized by the French philosopher René Descartes, who began by rejecting all ancient opinions, all the teachings that had been transmitted from the past and, with his mental slate wiped clean, to begin his thinking anew.

In the present instance, there has been the fascination of exploring long-standing mysteries. Are flying saucers for real? Was the Bermuda Triangle responsible for the loss of an inordinate number of ships and planes? Did a great sixth continent, Atlantis, disappear into the ocean, carrying with it an advanced civilization, sometime in the remote past? Did a strange monster inhabit, and does it perhaps still live in, Loch Ness? Does extrasensory perception exist in fact or is it a delusion? Scientists and philosophers have long sought answers to these and similar mysteries. Definitive solutions acceptable to all doubters are unlikely to be found. In most cases, a vast amount of literature, presenting various points of view, often contradictory, may be found.

The varied subjects dealt with in *Scientific Enigmas* are of long-time concern among historians and scientists—in some instances dating back thousands of years. They have provided themes for innumerable books and articles. For the most part, they remain mysteries, still generating controversy and widely differing viewpoints among experts.

A small library of noteworthy works includes attempts to find answers to some of these enigmas. There may be cited, for example, Clarke's *Mysterious World,* Evans's *Cults of Unreason,* Gardner's *Fads and Fallacies in the Name of Science*, Hering's *Foibles and Fallacies of Science*, Morgan's *Facts and Fallacies,* Sladek's *The New Apocrypha*, and the periodical *Skeptical Inquirer.* There are also numerous treatments of more limited topics, such as age of the earth, origin of man, Atlantis, Bermuda Triangle, Noah's Ark, flying saucers, discovery of the North Pole, discovery of the American continents, and extrasensory perception.

Scientific Enigmas has several objectives: first, to examine some celebrated examples of subjects that have long puzzled the world; second, to report on recent findings and opinions on these matters; and, third, to expose misleading information that has frequently led the general public and even the experts astray.

The author's primary intent has been to bring together a wide-ranging variety of matters of long-time interest to historians of science and to general readers fascinated by attempts to solve famous ancient mysteries. As noted, the opinions and writings of many leading authorities have been drawn upon to shed light on these topics. Because no final solutions are likely in most instances, theories and speculation will continue to proliferate.

The ignorance of so-called experts about scientific matters is revealed by Christopher Cerf and Victor Navasky in *The Experts Speak: The Definitive Compilation of Misinformation* (1984). The authors offer hundreds of examples to show that these "experts" have often been wrong about virtually every subject, including outer space, sex, the environment, politics, economics, the arts, sports, technology, history, science, and much else.

The apparently limited impact of modern science on "reasonably well-educated" persons is revealed by a survey of students at universities in Texas, California, and Connecticut, reported by Francis Harrold, professor of archeology at the University of Texas in Arlington. It was found that 15 percent of the students had faith in astrology, the ancient pseudoscience that believes an individual's fate is controlled by the position of the stars at the moment of birth. One-third of the students believed in UFOs, or invasion of aliens from outer space. Similar numbers accepted the lost continent of Atlantis, and the existence of "Bigfoot," a hairy monster from inner Asia and the Pacific Northwest. Harrold commented on his findings: "What really surprised us were the number of students who believed in what we call 'cult beliefs', or unproven pseudo-scientific theories. We all agree that for a leading scientific nation, this is not a good sign of the effectiveness of our science education."

I am indebted to Jane Bliss Downs for inspiring me to undertake the present work and providing useful ideas on its content and scope. As a daughter of a distinguished chemist, Robert N. Wilson, she has had a lifetime interest in scientific history and theory. I owe a particular debt to Richard L. Hay, Ralph Grim Professor of Geology, University of Illinois, whose expert criticism of the chapters on the age of the earth, beginning of life, and origin of man ensured greater scientific accuracy. Chapter 28, "Search for Medical Miracles," was reviewed for errors or mistaken conclusions by Dr. Lawrence L. Jeckel and Dr. William J. Keller, both prominent physicians.

The preparation of the manuscript of *Scientific Enigmas* was in the capable hands of Kirsten Kite and Sheryl Gocking. The index was compiled by Clara D. Keller. Assistance in assembling research materials was provided by Robert Perry.

CHAPTER

1

Taboos and Superstitions

In the course of surveying the history of science for the past 2,500 years, one is impressed continually by the influence of traditions, taboos, and superstitions. Generally speaking, these have had a retarding, even baneful, effect on scientific progress over the centuries. One might suppose that scientists would be immune or least susceptible of anyone to such prejudices and erroneous thinking, but frequently that has not been the case.

After the brilliant contributions made by such ancient writers as Hippocrates, Aristotle, Euclid, Archimedes, and Theophrastus, a long night descended. The Middle Ages were indeed the Dark Ages for science in general. Except for a few Arabic writers, there are no great names in all these centuries that can compare with those of the philosophers of ancient Greece and Alexandria. Progress in science was hopelessly held back by the stultifying atmosphere of medieval theology, religious dogma, and superstition. Christian thought became antagonistic to secular learning, identifying it with heathenism. During this interval of nearly a thousand years, such interest in science as existed took the form of philosophical discussion and inquiry rather than laboratory experiments or direct observations of nature.

Anything new or revolutionary was likely to be dangerous to the health of its authors. Self-preservation provided a strong motive not to depart from traditional tenets and beliefs—no matter how little based on demonstrated facts.

Tragic examples fill pages in science history. Roger Bacon in the thirteenth century was condemned by his Franciscan order for what it regarded as heretical doctrine, and spent his last years under house arrest. Nicolaus Copernicus was deterred for years from publication of his *On the Revolutions of the Celestial Spheres* because of the potential disapproval of the Church, and did not see the book until a few hours before his death. Giordano Bruno rejected the idea that the earth is the center of the universe and held that the universe is infinite, for which beliefs he was burned at the stake. The same fate was suffered by Michael Servetus, at the hands of John Calvin, for his heretical views, including a belief in the circulation of the blood. William Harvey was subjected to prolonged hostility by the University of Paris medical school faculty for his writings on blood circulation. Andreas Vesalius was roundly condemned because his findings on the structure of the human body ran contrary to the teaching of the ancient Galen, a second-century Greek physician, and he narrowly escaped trial by the Inquisition. Galileo was less fortunate; for his advocacy of the Copernican theories he was tried by the Inquisition and forced to spend his last years under house arrest. Controversies raging around Charles Darwin and the theory of evolution have persisted into the twentieth century, though burnings at the stake have gone out of fashion.

Let's consider these and other cases in more detail. Roger Bacon makes a good starting point. During his time in the thirteenth century, all Europe was cursed with a reverence for the dead. Aristotle was regarded as infallible, and to question Galen was heretical. Fixed dogmas about the heavens and the earth were passed on from generation to generation. It was thought that Aristotle had solved all problems in metaphysics and natural history, and that Galen had discovered all that could be known about the human body and medical science. Such science as was permitted to exist was based upon authority, revelation, and superstition. A sterile scholasticism dominated the universities.

Roger Bacon, probably the supreme intellect of his time, carried on a variety of experiments in his laboratory at Oxford. These experiments, plus his heretical teachings and outspoken criticisms of both ancient and contemporary scholars, soon created suspicion among his superiors in the Franciscan order. He was directed to cease his blasphemous teachings, then sent to Paris for ten years. He was kept in close confinement, denied all opportunity to write, and deprived of books and instruments. Bacon was allowed to return to Oxford for a few years, but was then sentenced by Pope Nicholas IV to imprisonment. The charge was "certain suspected heresies and novelties." Following conviction for these crimes, he remained in jail for fourteen years and lived only two years after being released.

Nicolaus Copernicus is another tragic example. The potential disapproval of the Church deterred him for years from publishing his masterpiece, *On the Revolutions of the Celestial Spheres.* The Protestant Revolution and the intellectual ferment of the Renaissance were making religious authorities extremely suspicious of unorthodox and revolutionary theories. Copernicus finally yielded to the entreaties of some of his admirers, and the book was printed in Germany. Soon afterward Copernicus suffered a stroke and died a few hours after the first copy was placed in his hands. Most contemporary opinion was in violent opposition to the Copernican system. Both Martin Luther and John Calvin condemned it. Not until 1615, however, was the book placed on the Catholic *Index Librorum Prohibitorum*, and there it remained for more than two centuries. The hostile reception accorded Copernicus's work is scarcely surprising. There was universal belief that the world had been created for man's express benefit and that the earth was the center of the universe, with the planets, stars, and sun revolving around it. Now, according to Copernicus, the earth was just one planet, and not even the largest. It is little wonder that Giordano Bruno was burned at the stake in 1600 for his active support of these ideas and that Galileo was tried by the Inquisition, forced to recant his belief that the sun is the central body about which the earth and other planets revolve, and condemned to house arrest for the last ten years of his life.

Another famous scientist who tangled with the Inquisition was Andreas Vesalius, one of the great anatomists of all time. His pioneer work, *On the Structure of the Human Body*, opened a new world. The effect of such a radical publication on a superstitious age was explosive. Because he denied concepts of the human body accepted for centuries and attacked the sacred writings of the ancient Galen, Vesalius was condemned uncompromisingly by the stubborn forces of tradition. Ecclesiastical authorities became alarmed by Vesalius's statement that man, although a descendant of Adam, did not have one less rib on one side than on the other, as stated in Genesis. He also destroyed the myth that somewhere in the vertebral column there was a "resurrection bone," from which the body was to be reconstructed after death. In 1563, Vesalius was convicted of heresy and condemned to death by the Inquisition. His patron, Philip II of Spain, commuted his sentence on condition that he make a pilgrimage to Jerusalem in expiation of his sins—an expedition on which he lost his life.

Much of the opposition to Vesalius's book arose from objections to the dissection of the human body. It is of interest to note that a century and a half later, William Cowper's *Anatomy of Human Bodies*, published at Oxford, aroused bitter controversy in England.

Nearly a century after Vesalius, William Harvey came up with his great discovery on the circulation of the blood. Again opposition arose from prejudice and conservatism. According to one contemporary writer, John Aubrey, some critics considered Harvey crackbrained. There was prolonged hostility on the part of the University of Paris medical school, and the professor of anatomy there persuaded the faculty to prohibit the teaching of Harvey's discovery. Fortunately, Harvey lived to see general acceptance of his theories. His election to the presidency of the College of Physicians, three years before his death, is evidence of his high standing among his professional colleagues.

In the following century, we have an instance of an attack on intellectual freedom in the first great general encyclopedia, the French *Encyclopédie* (1751-1772), edited by Denis Diderot. No other encyclopedia ever had such a stormy history. Since it embodied the essential spirit of the democratic and scientific movement in France immediately before the Revolution, bitter reaction from the entrenched forces of conservatism was inevitable. After the first two volumes appeared, the powerful ecclesiastics and the despotic French court forbade printing, reprinting, or sale of the *Encyclopédie*, on the grounds that it tended "to destroy royal authority, establish the spirit of independence and revolt, and lay a foundation for error, corruption, irreligion, and incredulity." To circumvent the ban, the remaining volumes were printed in secret, the names of author and editors were omitted, and a false imprint in Switzerland was used.

At the moment of apparent triumph, Diderot suffered a crowning blow when he discovered that Le Breton, the publisher, apprehending government disapproval, had been systematically censoring copy for the encyclopedia after Diderot had returned corrected page proofs to him. Though retaining its essential spirit, the work was damaged irreparably.

A few years later, the founder of the modern science of geology, James Hutton, ran into trouble for his pioneer book, *The Theory of the Earth.* Hutton was the first to realize the immensity of geological time. In terms of geology, recorded time is only yesterday. Such radical doctrine was too strong for most of the author's contemporaries to stomach. Hutton soon was charged with heresy for expounding ideas contrary to the Scriptures and was condemned as an enemy of religion. One critic pronounced Hutton's theories "not only hostile to sacred history, but equally hostile to the principles of probability and to the dictates of natural philosophy." One prominent ecclesiastic, Joseph Townsend, denounced Hutton for the sacrilegious notion that the earth's age should be measured in millions of years and for questioning that supernatural agents had caused various types of upheavals on the earth's surface.

Great medical pioneers also have had their share of troubles. An example is Edward Jenner, discoverer of the vaccination for smallpox. Immunization was obtained through inoculation with matter from cowpox.

Vociferous opposition to Jenner developed in medical and lay circles. Ministers denounced vaccination from their pulpits as man's interference with the ways of God. Antivaccinationist societies were formed, some of them still in existence. The fantastic idea was circulated that vaccinations would cause the human subject to sprout horns, grow a tail, low like a cow, bellow like a bull, and acquire other bovine characteristics.

Another series of controversies in the medical field arose in the nineteenth century over the relation of germs and lack of sanitation as causes of disease. Three names stand out: Dr. Oliver Wendell Holmes, fighting to stop the epidemic of deaths from puerperal or childbirth fever in the United States, Joseph Lister in England, and Louis Pasteur in France. Conditions prevailing in nineteenth-century hospitals were incredibly bad. A terrible stench pervaded rooms and corridors, caused by the suppuration of wounds with pus. A heavy death toll resulted from septicemia, a form of blood poisoning; pyemia, with abscesses spread throughout the body; hospital gangrene, characterized by mortification of the tissues; and erysipelas, a contagious skin disease. The death rate from operations ranged as high as 40 percent and was even higher for amputations. It was Joseph Lister's discovery of essential antiseptic principles and Louis Pasteur's discoveries of bacterial infections that dramatically reduced the death toll. Nevertheless, like all great trailblazers, Holmes in this country and Lister and Pasteur were attacked bitterly by other medical practitioners of their time, who were unwilling to concede any responsibility for spreading infection and death. All three lived to see their theories vindicated.

Probably the most controversial figure in the history of science is Charles Darwin. Contemporary reception of Darwin's celebrated book, *On the Origin of Species*, has been compared to "a conflagration like lightning in a full barn." If the revolutionary new theory were true, the biblical story of creation could no longer be accepted. Church authorities immediately viewed the Darwinian thesis as dangerous to religion and roused a storm of opposition. Though Darwin conceded that man is evolution's greatest triumph, the theory still made man a descendant from lower forms and a member of the animal kingdom. Inevitably, therefore, the Darwinian hypothesis was an object of bitter theological attack on the grounds that it contradicted the Bible and degraded man.

As late as the 1980s, a United States president, Ronald Reagan, expressed skepticism about evolution, in these words: "Evolution is theory only. In recent years it has been challenged in the world of science. If the evolutionary theory is going to be taught in the schools, then I would think that also the biblical theory of creation, which is not a theory but the biblical story of creation, should also be taught."

In response to Reagan's stance, the American Association for the Advancement of Science retorted that it could offer 100 million fossils to demonstrate the fact of evolution.

This litany on the trials and tribulations of scientists might be rounded out with consideration of the most-talked-about, and at the same time most-tabooed, subject in our society: sex. Specifically, we might look at Havelock Ellis and Alfred Kinsey. Ellis became the first notable English writer to discuss sex openly and with detachment. His monumental *Studies in the Psychology of Sex* represented many years of reading and research. The first of seven volumes was issued in London in 1897. Almost immediately the publisher, George Bedborough, was arrested and charged with "having unlawfully and wickedly published and sold, and caused to be procured and to be sold, a wicked, bawdy, and scandalous, and obscene book." Found guilty, Bedborough was released on condition that he destroy the stock and publish nothing further of a similar nature. Because of the savage stand taken by the English courts, Ellis determined to issue none of the succeeding volumes in England. Instead, the entire set was brought out by F. A. Davis Company of Philadelphia, with the sale limited to members of the legal, medical, and educational professions.

In the twentieth century, there has been a breakdown of Victorian conventions and interdictions, leading to an increased demand for objective data among physicians, psychiatrists, and persons concerned with such matters as sexual adjustments in marriage, the sexual guidance of children, and sex education.

Dr. Alfred Kinsey was a biologist whose previous research had been confined to studies of the gall wasp. One reviewer commented that "Kinsey has studied the sex behavior of the American male as though the American male, too, were a gall wasp." Courage was required to gather data for the study and later to publish the findings. The mere news that the investigation was in progress aroused violent opposition in some quarters. Why should a report of a scientific investigation by an eminent Indiana University zoologist stir up so much antagonism and cause so many attempts to suppress it even before publication? The primary reason, of course, was that it dealt with a taboo subject, and there was some fear of what the inquiry might reveal. In any case, with firm support from Indiana University and the Rockefeller Foundation's Medical Sciences Division, the original Kinsey report, *Sexual Behavior in the Human Male*, was published in 1953, followed by a series of more specialized studies.

Up to this point, scientific taboos have been discussed. Some attention now ought to be focused on scientific superstitions, or scientific errors or misconceptions. These have a long history, probably going back to primitive man. For present purposes, one might start with Aristotle.

Aristotle was the first great biologist, but he did have some curious notions. Some of these were due to the religious ban on the dissection of human bodies that prevailed among the Greeks and other ancient peoples. There was also a total lack of scientific instruments.

As a result, Aristotle reached such erroneous theories and conclusions as these: the heart is the seat of intelligence and the brain's function is to cool the heart; man has only eight ribs on each side; woman has fewer teeth than man; the male element in reproduction merely quickens and stimulates; plants do not reproduce sexually; the egg is not a living thing; there is no essential difference between arteries and veins; and female offspring are the result of failure of form to dominate matter, i.e., they are incomplete or imperfect males.

The greatest botanist and herbalist of ancient times was Theophrastus. In his book on the history of plants, Theophrastus tells of curious folklore and superstitions associated with plants. Although he was skeptical about most of them, he recorded them because he found them too interesting to omit. The root diggers' calling, he discovered, was governed by various strange beliefs, such as the following: if taboos were broken, the herbs collected lost their efficacy; it was necessary to eat garlic before going to gather certain plants; some plants had to be collected by day and others only by night; in given instances, the best quality of drugs could be obtained by approaching on the windward side; with others, top-quality material required the digger to face east while lifting the roots; one had to beware, too, of the woodpecker, buzzard hawk, and eagle, respectively, while plucking peony, feverwort, and black hellebore, taking the precaution of drawing circles around the plant with a sword or placing fruit and cakes in the hole from which the plant had been uprooted.

Pliny, in his *Natural History,* writing in the first century, offered hundreds of cures for different ailments. Here are two:

1. A tooth stops aching when worms are taken from a certain prickly plant, put with some bread in a pillbox, and bound on the arm on the same side of the body as the aching tooth.

2. Two bedbugs bound to the left arm in wool stolen from shepherds are a charm against nocturnal fevers; they are a charm against diurnal fevers if wrapped in a russet cloth instead.

Through his writings on medical botany and materia medica, Pliny influenced the progress of medicine for the next fifteen hundred years. It is a wonder that our ancestors survived.

Pliny's "cures" were no more bloodcurdling than some offered by the famous Swiss physician Paracelsus in the sixteenth century. A strange "weapon salve" was concocted by Paracelsus, the ingredients of which were moss grown on a skull exposed to the weather, human blood and human fat, a drug from a dried Egyptian mummy, armeniac, oil of roses, and linseed oil.

The state of medicine and chemistry in the eighteenth century may be judged from an episode in England, where in 1739 a woman chemist, Joanna Stephens, was awarded a prize of £5,000 for a remedy that was said to have "cured Mr. Walpole, the Prime Minister, who had been suffering from the stone." The ingredients included "eggshells and snails, a ball of soap, swine's crests burnt to a blackness, burdock seeds, and honey."

Early chemistry was riddled with fears, superstitions, and taboos, keeping it in an infant state. The medieval cult of alchemy, with its attachment to mysticism and the occult, still swayed even the most sophisticated minds. The alchemists refused to give up their obsession with the idea that baser metals could be transmuted into gold and silver.

Astronomy was held back by believers in astrology, some of whom we still have with us. Associated with that idea were superstitions about comets. The famous sixteenth-century Danish astronomer, Tycho Brahe, for example, wrote that comets "are formed by the ascending from the earth of human sins and weaknesses, formed into a gas, and ignited by the anger of God."

There were also many attempts to explain the origin of fossils. The most commonly held belief for a long time was that fossils were the remains of animals lost in Noah's Flood.

One of the most curious superstitions prevailing among early scientists was a belief that lower forms of life came into being by spontaneous generation, an idea that goes back at least to Aristotle, and was treated as an unquestioned fact by Ovid, Pliny, Lucretius, and Virgil. It was believed that living matter came from dead substances. Virgil tells of bees swarming to life from the body of a dead bull. As late as the seventeenth century, it was the popular belief that eels, salamanders, lizards, flies, bees, and other lower forms originated by spontaneous generation. An Italian scientist, Buonnani, taught that "a certain timberwood after rotting in the sea produced worms which engendered butterflies, and these butterflies became birds." This fallacy was not exploded until the invention of the microscope.

Fables about birds are of ancient origin and were treated seriously by early writers on the subject. It was believed, for example, that barnacles transform into geese, eagles can renew their youth, birds of paradise fly incessantly, there is a wool-bearing fowl, hummingbirds sleep for six months, and goats are milked by the nightjar or goatsucker. A fable going back to Aristotle, which held that some birds hibernate in the winter, did not die out until the end of the eighteenth century.

A strange theory about plants prevailed for a time among seventeenth-century botanists. They attempted to show a similarity between animal and plant functions. Two of the best-known men, Nehemiah Grew in England and Marcello Malpighi in Italy, believed that the sap in plants was pumped through the vessels of the wood by a sort of rhythmical pulsation, corresponding to the circulation of blood in animals. Since the plant has no heart, they invented the rhythmical squeezing of the wood vessels to take its place. Not long afterward, another biologist, Stephen Hale, showed that no such squeezing takes place and that there is no circulation in plants comparable to blood circulation in animals.

An optimistic conclusion to these remarks is that, despite the depressing influence of taboos and superstitions, science has continued its steady progress over the centuries. Although the pursuit of knowledge has been sometimes retarded, slowed down, and held back, there is no evidence that important ideas and discoveries have been lost. Truth triumphs in the end.

CHAPTER

2

Age of the Earth

The age of the earth has been debated heatedly since ancient times, generally with theologians on one side and men of science on the other. Martin Luther, citing Moses as his authority, concluded that the world could not have existed for more than six thousand years. His contemporary, Philip Melanchton, German scholar and religious reformer (1497-1560), was even more explicit, maintaining that "the world was created in 3963 B.C."

Also specific was a famous conclusion by James Ussher, Irish Archbishop of Armagh (1581-1656), who stated in his *Annals of the West*, "The world was created on 22nd October, 4004 B.C. at 6 o'clock in the evening." The Bishop's biblical chronology was slightly amplified and corrected by fifteen hours by Dr. John Lightfoot, vice-chancellor of the University of Cambridge. Using the same sources, Dr. Lightfoot in 1859 stated that "heaven and earth, center and circumference, were created together, in the same instant, and clouds full of water, by the Trinity on Sunday, twenty-first of October, 4004 B.C., at nine o'clock in the morning."

In the eighteenth century, a noted French scientist, Buffon, director of the Royal Museum of France, calculated that the earth would last 168,123 years, from the creation to the end of organic life.

In striking contrast to these relatively brief periods of time, research geologists have been steadily increasing their estimates of the age of the earth. Geologists have divided earth history into two eons: the first four billion years compose the Precambrian, where the earliest fossils (single-cell bacteria 3.1 billion years old) have been found; the remaining 600 million years, the Phanerozoic, mark the appearance of complicated life (worms and shellfish). The latter eon is divided into three *eras:* the Paleozoic or ancient life, the Mesozoic or middle life, and the Cenozoic or recent life. The Mesozoic is further divided into three *periods*: the Triassic, Jurassic, and Cretaceous.

Paleontologists have reconstructed more than 150 species, chiefly from the Jurassic and Cretaceous periods. The oldest have been found in Brazil. They lived about 240 million years ago. A mass extinction occurred 200 million years ago. Thereafter, the vertebrate world consisted mostly of early mammals, turtles, crocodiles, and small dinosaurs. All or nearly all dinosaurs disappeared at the end of the Cretaceous period. The latest thinking by scientists, trying to explain mass extinctions, concludes that asteroids collided with the earth to produce life-choking dust clouds. Some scientists conclude that there have been thirteen mass extinctions in the last 250 million years. It has been theorized that mass extinctions happened regularly every twenty-six million years.

Other efforts to explain the extinction of life on earth attribute the catastrophes to great floods. The most famous of these is the biblical story described in Genesis 6-9, of a deluge that overwhelmed the earth and destroyed all human life except Noah and his family. Deluge legends also occur in the folklore and early literature of other peoples. Ancient Babylonia had two stories, one traced to about 2000 B.C. Some scholars believe that the Genesis story is based on the Babylonian source, with which it has many parallels. Others believe that both are derived from a common Semitic source. Deluge stories are also found in India, Persia, Syria, and Greece.

Attempts to estimate the age of ancient objects have been facilitated and made more exact by the use of radiocarbon or carbon dating by anthropologists, geologists, and historians. Radiocarbon dating is useful and accurate on objects up to about forty thousand years old.

The significance of fossils was not recognized by early observers. During the Middle Ages and Renaissance, no one knew what to make of fossils. The most popular explanation was that they were relics of plants and animals destroyed by Noah's Flood. Many distinguished scholars believed that they had grown in the bowels of the earth by some sort of natural or occult process. Some argued that Satan had placed them there to mislead the faithful.

Summing up present views, Richard Hay, a leading geologist, and a member of the Louis Leakey team of British anthropologists in Tanzania, east Africa, comments, "The age of the earth is now accepted as 4.5 billion years. This is based on isotopic dating of meteorites, which are fragments of larger parent bodies that formed early in the history of the solar system. The age of meteorites was first established by Patterson in 1955-1956 on the basis of the isotopic composition of lead in three stone and two iron meteorites. Patterson also showed that terrestrial lead had the same isotopic composition as that of meteorites (more exactly it is in the range of meteorite lead), strongly supporting the common origin of the earth and meteorites. The 4.5 billion years age of meteorites is confirmed by other isotopic methods."

UPDATE

One of the main theories about the birth of the universe was expounded in 1930 by the Belgian astronomer George Lemaitre. This was the "Big Bang" theory, which suggested that about ten billion years ago all the matter of the universe was contained in a primal atom. The "cosmic egg" exploded and its many fragments became galaxies, one of which contains our solar system.

In May 1987 some of the world's most eminent scientists met to discuss quantum cosmology at Fermilab near Batavia. The group apparently accepted the validity of the Big Bang theory and proceeded to raise the question of what happened before and after. The time table proposed since creation of the universe was as follows: it has been fifteen billion years since the Big Bang, fourteen billion years since the galaxies were formed, and ten billion years since the solar system was formed.

CHAPTER

3

Origin of Life

The mystery of the beginning of life has intrigued philosophers, scientists, and theologians since time immemorial. Differences of opinion are irreconcilable and frequently bitterly fought. On one side are scientists who measure the age of the earth in billions of years and the origin of life perhaps a billion years ago. Opposing them are religious fundamentalists who accept literally the Genesis version set forth in the Bible, which provides for instant creation only a few thousand years ago.

A new dimension was added to the controversy in the mid-nineteenth century by Charles Darwin, British naturalist, who theorized that all animal species, including man, evolved from lower forms of life. If the revolutionary new theory were valid, the biblical theory of creation could no longer be accepted. The Church immediately saw the Darwinian theories as dangerous to religion, and roused a storm of opposition. Even such a noted scientist as Louis Agassiz stated, "I trust to outlive this mania."

The basic principles of Darwin's theories are today almost universally accepted in the scientific world. Nevertheless, controversies have raged around them for more than a hundred years. The fundamentalist-modernist battles of the mid-twenties, culminating in the Scopes "monkey trial" in Tennessee, are among the most publicized instances of a war that began in the 1850s. William Jennings Bryan, who led the antievolution forces at the Dayton trial, declared, "All the ills from which America suffers can be traced back to the teaching of evolution. It would be better to destroy every other book ever written, and save just the first three verses of Genesis."

There have been no signs of an armistice in the hostilities. Much of the action has shifted to state legislatures, which have been asked to ban the teaching of evolution in the schools or to require that equal time be provided for the theory of instant creation whenever evolution is taught. In a 1987 decision, the U.S. Supreme Court declared unconstitutional a Louisiana law requiring equal time be devoted to the teaching of creationism and evolution. The Louisiana law was rejected on the ground that it promoted certain religious beliefs and was therefore in violation of the principle of separation of church and state.

Despite various cases of condemnation, Darwin's impress on nearly all major fields of learning continues to be profound, viewed in retrospect. The doctrine of organic evolution has been accepted by biologists, geologists, chemists, and physicists, by anthropologists, psychologists, educators, philosophers, and sociologists, as well as historians, political scientists, and philologists.

One of the most bizarre theories on the origin of life, the concept of spontaneous generation, which supposed that living things were produced from dead matter, was referred to in the introduction. Aristotle wrote that maggots came from dead meat, fleas from dust, etc. Centuries later, Pasteur demonstrated that bacteria could be produced only by other living bacteria, effectively disproving the idea of spontaneous generation.

A famous example of efforts to disprove evolutionary teaching was Philip Gosse's book *Omphalos* (1857), which overcame all conflict between evolution and the Bible by explaining that God created the earth in six days about 4004 B.C., and then cunningly covered his tracks by forging elaborate geological and paleontological records that could lead misguided scientists into believing all sorts of nonsense like natural selection or the immense age of the earth.

There can be no doubt that there was a long period in the history of the universe when no life existed on the earth. Temperatures were too hot for any living creature to survive. But at least three billion years ago or more, the surface of the earth became cool enough for living creatures to survive. A simple form of life from that period, bacteria, has been identified.

A crucial question then arises: how did life come to be upon the earth? A great historian of science, Sir Arthur Thomson, in his *Riddles of Science* (1932), answers, "The problem is too difficult and the clues are so vague that we can well understand those who shake their heads and refuse to express an opinion. One hypothesis is that germs of life may have reached the cooled earth from outer space. If so, the dilemma remains: from what planet did they come and how did they arise in their original home?"

In a diagram of geological ages and major events in evolution, Thomson concludes that "from the emergence of man to the appearance of the first distinct animals in the Cambrian, could not be less than 500 million years." The first birds appeared in the Jurassic period. Reptiles were dominant in the Mesozoic ages; the first mammals probably emerged in the Triassic era; reptiles dominated the Carboniferous period, fishes were abundant in the Silurian period, and many invertebrate groups in the Ordovician.

Life appeared more than four billion years ago. By 500 million years ago, sponges, sea cucumbers, worms, brachiopods, trilobites, and crustaceans lived in the sea. Fishes, which began over 450 million years ago, were well established by the Silurian period, which began about 430 million years ago. Amphibians developed in the Devonian period, which began about 410 million years ago. Reptiles appeared in the Carboniferous or Pennsylvanian period beginning 360 million years ago. During the Jurassic period, beginning 210 million years ago, flying reptiles and birds appeared. The earliest-known bird is the archeopteryx. Mammals evolved from reptiles during the Jurassic period. Mammals became the dominant form of life sixty-five million years ago. A piglike animal, a three-toed horse, a two-horned rhinoceros, and a long-necked camel are known to have existed twenty million years ago. Man, the most successful mammal, has ruled the earth for less than 100,000 years.

The International Society for the Study of the Origins of Life in Washington, D.C., the Center for the Study of Evolution and the Origin of Life at the University of California at Los Angeles, and the University of Maryland Laboratory of Chemical Evolution are devoted to investigating the origin of life. An international meeting on this subject was called at Berkeley, California, in the summer of 1986. There was a consensus, generally agreed to by the participants, that the creation of life was not a chance occurrence but the end product of a lengthy process of chemical evolution that occurs commonly throughout the universe. As one scientist stated, "The universe appears to be in the business of making life."

No one is yet certain, however, how or where life actually began on earth. The main difficulty is that the earliest geologic record of the earth, formed some 4.6 billion years ago, has been erased. The oldest rocks capable of preserving fossils that have been recovered by scientists are 3.8 billion years old, a time when bacteria were already flourishing and beginning to diversify. Most of the older rocks are believed to have been destroyed in showers of meteors and comets that pelted the earth's surface.

Some scientists believe life's birthplace was in hot, deep ocean vents, where gases such as methane and ammonia accumulate. Others insist that it happened in shallow tidal pools, or in glaciers, or in a pressurized atmosphere laden with water vapor, or in mineral-rich clays.

There was unanimous agreement among the scientists gathered at Berkeley that life started on the earth. They dismissed the idea of anything being "blown in from outer space." Further, they believed that life started in the earliest geologic period, the Archean. Bacteria appeared on the earth two billion years before any other creatures. Cells with nuclei, which make up the rest of the living world, appeared 1.5 billion years ago. During the last 500 million years of the earth's history, these cells developed into plants and animals, the scientists concluded.

The more complex life forms are thought to have lived in the sea. Eventually plants and later animals developed that were able to live on land. The first land animals were probably scorpions. Insects and spiders followed, then amphibians, and still later reptiles.

CHAPTER

4

Dinosaurs

A reference to dinosaurs was made in the previous chapter on the origin of life. The subject deserves further consideration. The first dinosaur bones were discovered in the early 1800s along the Dorset coast in England. During the past century and a half, specimens have been found in every part of the world except arctic zones. Dinosaur remains have been particularly abundant in the western United States and in Tanzania, east Africa.

Dinosaurs are a long-extinct group of reptiles that for millions of years were the dominant land animals of the Mesozoic era. The name *dinosaur* comes from the Greek words meaning "terrible lizard." The group comprised a great number of different kinds of reptiles. About 250 genera have been named.

The popular conception of dinosaurs is of creatures of immense size, and in fact dinosaurs as a whole tended to be of gigantic proportions, the largest animals that ever walked the earth. Some reached lengths of eighty feet or more and probably weighed forty to fifty tons. Some species, however, remained small, weighing little more than a rabbit. Certain of the dinosaurs were flesh-eaters, but a majority lived on a herbivorous diet.

A graphic description of one genus of sauropods, a brontosaurus-like dinosaur, was offered by paleontologist Walter P. Coombs, Jr.: "A creature with the nose of an extinct South American mammal, the neck of a giraffe, the limbs of an elephant, the feet of an extinct relative of the horse, the lungs of a bird and the tail of a lizard."

Dinosaurs are classified in two orders: Ornithischia (bird-hip) and Saurischia (lizard-hip). An important difference between the two groups is the structure of the pelvis, which in the first class is four-pronged and in the second three-pronged.

Dinosaurs lived from the late Triassic period, some 200 million years ago, through the Jurassic period, lasting to 140 million years ago, until the end of the Cretaceous period, ending about sixty-three million years ago.

Among the saurischian dinosaurs were the brontosaurus, largest of dinosaurs with long necks and tails and small heads. One of the flesh-eaters was tyrannosaurus rex, a two-legged creature with long tail, short fore-limbs, huge head, and menacing teeth. This beast was nearly fifty feet long, and in a walking position was approximately twenty feet high. One of the most terrifying animals ever to walk the earth, it was believed to have subsisted mainly on other dinosaurs.

Also among the great variety of dinosaurs were duck-billed, plant-eating dinosaurs, generally living in the water along river and lake shores; horned dinosaurs, also herbivorous, four-footed, with horns on the head; armored dinosaurs, with broad skull and strongly protected by armor plates; plated dinosaurs, four-footed armored types; carnivorous dinosaurs, primarily flesh-eaters, that walked on hind legs and used front limbs for grasping; and sauropods, herbivorous dinosaurs that generally walked on all fours, the giants of the dinosaur world.

Contemporary with dinosaurs was the first known bird, archeopteryx. This creature had wings and was probably able to fly, but in many details it was essentially reptilian. In early Cenozoic times there was considerable development of large, flightless birds similar to present-day ostriches and rheas. A theory that has received considerable attention is that birds are direct descendants of one type of dinosaurs—a theory generally rejected by paleontologists. Despite certain striking similarities, the accepted view is that birds and dinosaurs developed independently of each other.

It is certain that no human being ever saw a live dinosaur. The last of the dinosaurs became extinct about sixty-three million years ago. The human race, according to anthropologists, has lived for only a couple of million years and civilized man's recorded history goes back only a few thousand years.

One of the great unresolved enigmas of science is how to account for the sudden and complete extinction of dinosaurs. A question often asked is, "Why did the dinosaurs die out?" No generally accepted answer has been found. In 1963 a Princeton University professor, Glenn Jepsen, listed forty-six theories. Geological changes may have been a major factor in the decline of the dinosaurs. New mountain chains, such as the Rockies, were rising, which caused climatic and other environmental changes disadvantageous to the dinosaurs. The world's flora changed considerably at the end of the Cretaceous times, and many of the herbivorous dinosaurs were unable to adapt themselves to new types of vegetation as a food supply. Mammals were rapidly evolving with larger brains, and these warm-blooded animals were too much competition for the sluggish, cold-blooded dinosaurs. Winters gradually became colder, and dinosaurs lacked protection against the cold, for they had neither fur nor feathers.

Further speculation has suggested that an overabundance of meat-eating dinosaurs ate all the plant-eaters and then themselves died of hunger; that little mammals were fond of dinosaur eggs and ate so many that none were left for purposes of reproduction; that the dinosaurs were poisoned by the water they drank or the plants they ate. Many other causes have been put forward to explain the dinosaurs' disappearance. The simple fact is that the real facts are unknown. The problem of their extinction is likely to always remain as one of the great unsolved mysteries of science.

REFERENCES

Allaby, Michael, and James Lovelock. *The Great Extinction; The Solution to One of the Great Mysteries of Science: The Disappearance of the Dinosaurs.* Garden City, N.Y.: Doubleday, 1983.

Bakker, Robert. *The Dinosaur Heresies.* New York: Morrow, 1987.

Blough, Glen O. *Discovering Dinosaurs.* New York: McGraw-Hill, 1960.

Charig, Alan. *A New Look at the Dinosaurs.* New York: Facts on File, 1979.

Colbert, Edwin H. *Age of Reptiles.* New York: Norton, 1965.

Colbert, Edwin H. *The Dinosaur Book: The Ruling Reptiles and Their Relatives.* New York: American Museum of Natural History, 1945.

De Camp, L. Sprague. *The Day of the Dinosaur.* Garden City, N.Y.: Doubleday, 1968.

Dunkle, David H. *World of the Dinosaurs.* Washington, D.C.: n.p., 1966.

Kurten, Bjorn. *The Age of Dinosaurs.* New York: McGraw-Hill, 1968.

Ostrom, John. *Strange World of Dinosaurs.* New York: Putnam, 1964.

Stout, William, and others. *The Dinosaurs.* New York: Bantam Books, 1981.

Swinton, William E. *The Dinosaurs.* New York: Wiley, 1970.

CHAPTER

5

Homo Sapiens

The present chapter is restricted to a consideration of the development of man at the stages designated as *Homo erectus* and *Homo sapiens*, omitting the earlier, more primitive evolutionary levels dealt with in the essay on the origin of life.

In his *On the Origin of Species*, Charles Darwin intentionally soft-pedaled any discussion of man's beginning, recognizing that any emphasis on this phase of evolution would cause his entire theory to be rejected. Within his broad structure of the theory of evolution, therefore, Darwin did not include the human animal. Any application of the theory to account for the emergence and development of man was omitted. There were few references to apes and monkeys, and only one sentence concerning the author's own species. "Light will be thrown on the origin of man and his history." Despite this cautious approach, it was obvious to any intelligent reader that the theory and principle of evolution could easily be extended to include the human zoological group. Bitter controversy immediately erupted, with Darwin coming under attack not only by religious conservatives but by many scientists and philosophers. There were attempts to discredit Darwinian evolution by ridiculing similarities between man and the great apes, gorillas, and chimpanzees.

Darwin answered his critics in 1871 with publication of *The Descent of Man*. Therein a massive amount of evidence was advanced to demonstrate that the human race is also a product of evolution from lower forms. In an eloquent introductory statement, Darwin explains: "We must, however, acknowledge, as it seems to me, that man with all his noble qualities, with sympathy which feels for the most debased, with benevolence which extends not only to other men but to the humblest living creature, with his god-like intellect which has penetrated into the movements and constitution of the solar system—with all these exalted powers—Man still bears in his bodily frame the indelible stamp of his lowly origin."

Was Darwin justified in concluding that man sprang from a stock common to the anthropoid apes and to himself? His conclusion has been corroborated by all that has been discovered since 1871, when *The Descent of Man* was published.

The earliest-known hominids (manlike creatures) lived about 5.5 million years ago. The planet earth once belonged to the invertebrates, followed by the fishes, and the reptiles, and finally by the human species. How did man achieve this dominant place? The human animal is believed to have emerged in central eastern, northern, and south Africa. Hominid footprints 3.5 millions years old are evidence of the emergence and survival of man's oldest ancestors, the dawn of mankind on the earth. By then, early man had achieved a bipedal human gait although his brain had not yet reached a size approaching that of modern man.

Manlike creatures and men are biologically classified in the family Hominidae. The *Homo* genus includes four major groups of man: *Homo habilis*, earliest man; *Homo erectus*, early man; *Homo sapiens*, Neanderthal man; and *Homo sapiens sapiens*, modern man. The earliest-known manlike creatures, the *Australopithecines*, lived between about 5.5 million and one million years ago. They walked nearly erect and had nearly human teeth. Their brains were relatively small, with apelike jaws (giving them an apelike appearance), and their skull capacity (600 cubic centimeters, compared to about 1,350 cubic centimeters in modern man) limited their intelligence. The *Australopithecines* probably used animal bones and teeth for tools.

Homo habilis (1.3 million to two million years ago) is widely accepted as the earliest man. He is much smaller than *Homo erectus*, and almost certainly used stone tools. *Homo erectus* perhaps emerged as long ago as 1.6 million years and existed until about 200,000 years ago. This species was of medium height (five feet to five feet eight inches), walked erect, with arms and legs like those of modern man. His head was more primitive in appearance, with large jaws, heavy brow ridges, retreating forehead, and skull capacity of about 1,000 cubic centimeters. *Homo erectus* lived in Java, China, northern and eastern Africa, and probably in Europe. There is evidence that he used fire, made stone tools, and hunted animals.

The best-known fossil remains of *Homo erectus* include:

1. Heidelberg man, the earliest form of primitive man in Europe, based on a jawbone found near Heidelberg, Germany, in 1907. The age is probably between 250,000 and 500,000 years.

2. Java man, an early form of man living during the Pleistocene epoch, 500,000 or more years ago, found on the island of Java in 1891 and again in 1936-1939. Java man was about five feet eight inches tall, walked erect, with a brain about three-fourths as large as modern man's. He had prominent brow ridges, a retreating forehead, and large powerful jaws.

3. Peking man, who lived more than 300,000 years ago, based on the remains of forty-five individuals found in a cave near Peking, China. Peking man stood erect, made stone tools, and used fire. His skull and teeth resemble those found in modern Mongoloid racial groups.

The next stage of human development is Neanderthal man, who lived during the latter part of the Glacial period, probably 100,000 to 40,000 years ago. Skeletons have been found in Europe, northern Africa, Asia Minor, and central Asia. The remains show some of the characteristics of modern man. Neanderthal man was about five feet tall, stout-bodied, and very muscular. He had large hands and feet, a broad face and prominent nose, large teeth, and massive jaws. He stood fully erect. Neanderthal man was a hunter and cave dweller who made chipped-stone tools and weapons. Why he disappeared is not known. In Europe he was replaced by modern men of the Cro-Magnon type.

Cro-Magnon man was one of the earliest types of modern man. The first examples were found in France in 1868. Later, similar skeletons were discovered elsewhere in Europe, Asia Minor, and Central Asia. Cro-Magnon man lived toward the end of the last Ice Age, about 20,000 to 50,000 years ago, replacing the more primitive Neanderthal man. The Cro-Magnons were tall, with high foreheads, broad faces, and prominent chins. They made weapons and tools of bone and flint, and jewelry of shells and ivory. On the walls of caves that they inhabited they drew colored pictures of the animals that they hunted.

The emergence of Cro-Magnon man preceded the appearance of modern man (both belonging to the species *Homo sapiens sapiens*). A hunting-gathering life-style was gradually replaced by the coming of agriculture and the cultivation of plants, the domestication of animals, metallurgy, and finally civilization.

Throughout the entire sixty-five million years of the current Cenozoic era, the primate order has been undergoing successful evolution. The human species is directly related to, although not descended from, the three great apes (orangutan, chimpanzee, and gorilla). The other primates are primarily arboreal (tree-dwelling) creatures. The human being has adapted to a terrestrial existence. In that process it has acquired an erect posture and the ability to make and use a variety of instruments. A central nervous system, including the cerebral cortex, has enabled the human to develop verbal language and other tools essential to a high social order.

Anthropologists have attempted to explain the differences among major racial groups. Some believe that variations may go back to the *Homo erectus* stage, later evolving separately into modern types. In his *The Origin of Races*, Carleton S. Coon classifies the living peoples of the world into five basically geographical groups: the Caucasoid (includes Europeans and their overseas kinsmen, the Middle Eastern whites from Morocco to West Pakistan, and most of the people of India, as well as the Ainu of Japan); the Mongoloid (most of the East Asiatics, Indonesians, Polynesians, Micronesians, American Indians, and Eskimo); the Australoid (Australian aborigines, Melanesians, Papuans, some of the tribal folk of India, and the various Negritos of South Asia and Oceania); Congoid (Negroes and Pygmies of Africa); and Capoid (Bushmen and Hottentots, and other relict tribes, like the Sawdawe of Tanganyika). Coon concludes that "all the evidence from comparative ethnology, linguistics, and prehistoric archeology indicates a long separation of the principal races of man."

Leaders in exploring Africa for human beginnings were the Leakeys: Louis S. B. Leakey, British anthropologist and paleontologist, his wife, Mary, and son, Richard. Louis's early archeological discoveries pointed to Africa as the site of the origin of man. Among Mary's significant finds was a Miocene form possibly ancestral to both man and ape. In 1959, at Olduvai Gorge in Tanzania, the Leakeys found *Australopithecus* fossils and *Homo habilis* remains that were 1.8 million years old. Mary, herself a distinguished archeologist, discovered in the same area what were at the time the oldest-known hominid fossils—teeth and jaws 3.5 to 3.75 million years old. Richard, excavating near Lake Rudolphi in Kenya from 1969 to 1972, found magnificent skulls of an *Australopithecus* and what he identified as *Homo habilis* and *Homo erectus* fossils that are 1.3 to 1.8 million years old.

Now numbering some five billion individuals worldwide, the human race has spread throughout the globe from the tropics to the arctic regions. Man has shown amazing adaptability to any climate, to every geographical condition. All other living creatures must serve him or risk extinction. Man's prolific growth, however, places him in danger of exhausting the earth's resources and exceeding the ability of his universe to support him.

REFERENCES

Boule, Marcellin. *Fossil Men.* New York: Dryden Press, 1957.

Brown, Frank, and others. "Early Homo Erectus Skeleton from West Lake Turkana, Kenya." *Nature* 316 (August 1985): 788-92.

Leakey, L. S. B. *Adam's Ancestors.* New York: Harper, 1960.

Leakey, L. S. B. *The Progress and Evolution of Man in Africa.* New York: Oxford University Press, 1963.

Leakey, L. S. B. *Unveiling Man's Origins.* London: Methuen, 1970.

Leakey, L. S. B., and Mary D. Leakey. "Recent Discoveries of Fossil Hominids in Tanganyika at Olduvai and Near Lake Natron." *Nature* 202 (4 April 1964): 3-4.

Leakey, L. S. B., P. V. Tobias, and J. R. Napier. "A New Species of the Genus Homo from Olduvai Gorge." *Nature* 202 (4 April 1964): 7-9.

Leakey, Richard E. *People of the Lake.* Garden City, N.Y.: Anchor Press, 1978.

Leakey, Richard E., and Roger Lewin. *Origins.* New York: Dutton, 1979.

Montagu, Ashley. *Statement on Race.* New York: Oxford University Press, 1972. (Includes UNESCO statements on race.)

Susman, R. L., and J. T. Stern. "Functional Morphology of Homo Habilis." *Science* 217 (September 1982): 931-34.

Tobias, P. V. "The Olduvai Bed I *Hominine* with Special Reference to Its Cranial Capacity." *Nature* 202 (4 April 1964): 3-4.

CHAPTER

6

Races of Mankind

The previous chapter summarized what is known or is believed by scientists about the origin of *Homo sapiens* or man. Equally complex is the question of races of man.

All human beings belong to the same species, *Homo sapiens*. There are more similarities between races than differences. All the many kinds of humans are believed to be descended from the same early group. All human beings and groups are closely related. Groups are classified according to obvious physical traits, such as color of skin, hair, and eyes; form of hair, whether straight or curly; shape of the head and facial features; and body size.

There are many mistaken ideas about race. Racial traits are physical and should not be confused with nationality, religion, or language. There is, for example, no "Swedish race," although a majority of Swedish people belong to the Nordic division of the Caucasoid race. Biologically speaking, there is no Aryan race, although the Nazis of Germany claimed that they were keeping such a "race" pure by exterminating the "inferior race" of Jews. Actually, there is no "Jewish race," although most adherents of Judaism are descended from tribes that spoke the same language. Similarly, there is no "English race" and no "Irish race"; these are political groups. The so-called white or European race is in fact a collection of race populations, belonging to a large geographical area.

Nevertheless, it has long been traditional for one group of people to boast of its superiority over other groups, for economic or political purposes. Racial discrimination has existed at one time or another in nearly every nation. At present, South Africa is one of the few nations to openly practice total racial segregation with the aim of perpetuating white supremacy.

Peoples of northwestern European origin have spread over the world's continents and number close to half a billion. Realistically, it could not be claimed that they belong to one race. The few genuine races, in contrast, are only a fraction of this size. Among them are the Ainu of northern Japan, the Bushmen of the Kalahari Desert in South Africa (numbering about 25,000), and the Lapps of Norway, Sweden, and Finland.

Races are continually changing. Existing races are no more than a small percentage of the total number of races that have existed in the course of human development.

Geographical races have been created by such factors as continental isolation, oceans, mountains, deserts, or other natural barriers separating races. The Sahara, for example, was the original southern limit of the European geographical race, and the northern limit of the African geographical race. The Himalayas arc the boundary between the Asian and Indian geographical races. The North Pacific and Arctic oceans separate the Asian and American Indian geographical races.

Students of race have investigated racial differences that seem to be of genetic origin rather than coming from changes in nutrition or way of life. An obvious racial difference is skin pigmentation. Color of the hair and eyes is also considered. Hair, for example, may be blond, red, dark, curly, or frizzy. Facial features may differ in nose form, lip thickness, and size of teeth. Blood-type classification systems have become increasingly important.

Considerable controversy has risen on the use of intelligence tests for measuring mental ability among different races. Critics point out the application of standard intelligence tests is questionable because they are not measures of innate ability and do not take into account such factors as parental education, parental attitudes, social class, personality, and home environment. To be valid, the tests should measure racial differences in verbal ability, abstract reasoning, and such skills as tonal memory and form identification. The racist arguments of the nineteenth century were based primarily on the measurement of human skulls. This approach has been totally discredited and replaced by intelligence tests which are now under attack.

A reasonably objective view on questions of race was taken in two statements issued by the United Nations Educational, Scientific and Cultural Organization. The first, issued in 1950, was largely the work of a group of social scientists. The second was the product of a group of physical anthropologists and geneticists. The two statements are remarkably similar in their conclusions.

Following is a portion of the views of the UNESCO social scientists on questions of race: First, "scientists have reached general agreement in recognizing that mankind is one: that all men belong to the same species, *Homo sapiens*. It is further generally agreed among scientists that all men are probably derived from the same common stock; and that such differences as exist between different groups of mankind are due to the operation of evolutionary factors of differentiation such as isolation, hybridization, and natural selection."

Second, "from the biological standpoint, the species *Homo sapiens* is made up of a number of populations, each one of which differs from the others in the frequency of one or more genes. Such genes, responsible for the hereditary differences between men, are always few when compared to the whole genetic constitution of man and to the vast numbers of genes common to all human beings regardless of the population to which they belong. This means that the likenesses among men are far greater than their differences."

Third, "a race, from the biological standpoint, may therefore be defined as one of the group of populations constituting the species *Homo sapiens*. These populations are capable of interbreeding with one another but, by virtue of the isolating barriers that in the past kept them more or less separated, exhibit certain physical differences as a result of their somewhat different biological histories. These represent variations, as it were, on a common theme."

"Whatever classification the anthropologist makes of man," the UNESCO social scientists point out, "*he never includes mental characteristics as part of those classifications.* It is now generally recognized that *intelligence tests do not in themselves enable us to differentiate safely between what is due to innate capacity and what is the result of environmental influences,* training, and education. So far as temperament is concerned, there is no definite evidence that there exist inborn differences between human groups. There is evidence that whatever group differences there might be are greatly overridden by individual differences and by differences springing from environmental factors. ... As for personality and character, these may be considered raceless. In every human group a rich variety of personality and character types may be found, and there is no reason for believing that any human group is richer than any other in these respects. With respect to race-mixture, the evidence points unequivocally to the fact that this has been going on from the earliest times."

In summary, UNESCO social scientists conclude that "the biological fact of race and the myth of race should be distinguished. For all practical social purposes, 'race' is not so much a biological phenomenon as a social myth. This myth has created an enormous amount of human and social damage. In recent years it has taken a heavy toll in human lives and caused untold suffering. It still prevents the normal development of millions of human beings and deprives civilization of the effective cooperation of productive minds. The biological differences between ethnic groups should be disregarded from the standpoint of social acceptance and social action. Unity of mankind from both the biological and social viewpoints is the main issue. To recognize this and to act accordingly is the first requirement of modern man."

A great biologist wrote in 1871: "As man advances in civilization, and small tribes are united into larger communities, the simplest reason would tell each individual that he ought to extend his social instincts and sympathies to all the members of the same nation, though personally unknown to him. This point being once reached, there is only an artificial barrier to prevent his sympathies extending to the men of all nations and races." These are the words of Charles Darwin in *The Descent of Man.*

Two years after their initial UNESCO statement, a group of physical anthropologists and geneticists issued a separate statement outlining their general principles:

> We now have to consider the bearing of these statements on the problem of human equality. It must be asserted with the utmost emphasis that equality as an ethical principle in no way depends upon the assertion that human beings are in fact equal in endowment. Nevertheless, the characteristics in which human groups differ from one another are often exaggerated and used as a basis for questioning the validity of equality in the ethical sense. For this purpose we have thought it worth while to set out in a formal manner what is at present scientifically established concerning individual and group differences.
>
> 1. In matters of race, the only characteristics which anthropologists can effectively use as a basis for classification are physical and physiological.

2. According to present knowledge there is no proof that the groups of mankind differ in their innate mental characteristics, whether in respect to intelligence or temperament. The scientific evidence indicates that the range of mental capacities in all ethnic groups is much the same.

3. Historical and sociological studies support the view that genetic differences are not of importance in determining the social and cultural differences between different groups of *Homo sapiens*, and that the social and cultural *changes* in different groups have, in the main, been independent of *changes* in inborn constitution. Vast social changes have occurred which were not in any way connected with changes in racial type.

4. There is no evidence that race mixture as such produces bad results from the biological point of view. The social results of race mixture whether for good or ill are to be traced to social factors.

The conclusions of these UNESCO physical anthropologists and geneticists were as follows:

1. Scientists are generally agreed that all men living today belong to a single species, *Homo sapiens*, and are derived from a common stock, even though there is some dispute as to when and how different human groups diverged from this common stock.

   The concept of race is unanimously regarded by anthropologists as a classificatory device providing a zoological frame within which the various groups of mankind may be arranged and by means of which studies of evolutionary processes can be facilitated. In its anthropological sense, the word *race* should be reserved for groups of mankind possessing well-developed and primarily heritable physical differences from other groups. Many populations can be so classified but, because of the complexity of human history, there are also many populations which cannot easily be fitted into a racial classification.

2. Some of the physical differences between human groups are due to differences in hereditary constitution and some to differences in the environments in which they have been brought up. In most cases, both influences have been at work. The science of genetics suggests that the hereditary differences among populations of a single species are the results of the action of two sets of processes. On the one hand, the genetic composition of isolated populations is constantly but gradually being altered by natural selection and by occasional changes (mutations) in the material particles (genes) which control heredity. Populations are also affected by fortuitous changes in gene frequency and by marriage customs. On the other hand, crossing is constantly breaking down the differentiations so set up. The new mixed populations, in so far as they, in turn, become isolated, are subject to the same processes, and these may lead to further changes. Existing races are merely the result, considered at a particular moment in time, of the total effect of such processes on the human species. The hereditary characters to be used in the classification of human groups, the limits of their variation within these groups, and thus the extent of the classificatory subdivisions adopted may legitimately differ according to the scientific purpose in view.

3. National, religious, geographical, linguistic and cultural groups do not necessarily coincide with racial groups; and the cultural traits of such groups have no demonstrated genetic connection with racial traits. Americans are not a race, nor are Frenchmen, nor Germans; nor *ipso facto* is any other national group. Moslems and Jews are no more races than are Roman Catholics and Protestants; nor are people who live in Iceland or Britain or India, or who speak English or any other language or who are culturally Turkish or Chinese and the like, thereby describable as races. The use of the term *race* in speaking of such groups may be a serious error, but is one which is habitually committed.

4. Human races can be, and have been differently classified by different anthropologists. Most of them agree in classifying the greater part of existing mankind into at least three large units, which may be called major groups (in French *grand races*, in German *Hauptrassen*). Such a classification does not depend on any single physical character, nor does, for example, skin color by itself necessarily distinguish one major group from another. Furthermore, so far as it has been possible to analyze them, the differences in physical structure which distinguish one major group from another give no support to popular notions of any general "superiority" or "inferiority" which are sometimes implied in referring to these groups.

   Broadly speaking, individuals belonging to different major groups of mankind are distinguishable by virtue of their physical characters, but individual members, or small groups, belonging to different races within the same major group are not so distinguishable. Even the major groups grade into each other, and the physical traits by which they and the races within them are characterized overlap considerably. With respect to most, if not all, measuring characters, the differences among individuals belonging to the same race are greater than the differences that occur between the observed averages for two or more races within the same major group.

5. Most anthropologists do not include mental characteristics in their classification of human races. Studies within a single race have shown that both innate capacity and environmental opportunity determine the results of tests of intelligence and temperament, though their relative importance is disputed.

   When intelligence tests, even non-verbal, are made on a group of non-literate people, their scores are usually lower than those of more civilized people. It has been recorded that different groups of the same race occupying similarly high levels of civilization may yield considerable differences in intelligence tests. When, however, the two groups have been brought up from childhood in similar environments, the differences are usually very slight. Moreover, there is good evidence that, given similar opportunities, the average performance (that is to say, the performance of the individual who is representative because he is surpassed by as many as he surpasses), and the variation round it, do not differ appreciably from one race to another.

Even those psychologists who claim to have found the greatest differences in intelligence between groups of different racial origin, and have contended that they are hereditary, always report that some members of the group of inferior performance surpass not merely the lowest ranking member of the superior group, but also the average of its members. In any case, it has never been possible to separate members of two groups on the basis of mental capacity, as they can often be separated on a basis of religion, skin color, hair form or language. It is possible, though not proved, that some types of innate capacity for intellectual and emotional responses are commoner in one human group than in another, but it is certain that, within a single group, innate capacities vary as much as, if not more than, they do between different groups.

The study of the heredity of psychological characteristics is beset with difficulties. We know that certain mental diseases and defects are transmitted from one generation to the next, but we are less familiar with the part played by heredity in the mental life of normal individuals. The normal individual, irrespective of race, is essentially educable. It follows that his intellectual and moral life is largely conditioned by his training and by his physical and social environment.

It often happens that a national group may appear to be characterized by particular psychological attributes. The superficial view would be that this is due to race. Scientifically, however, we realize that any common psychological attribute is more likely to be due to a common historical and social background, and that such attributes may obscure the fact that, within different populations consisting of many human types, one will find approximately the same range of temperament and intelligence.

6. The scientific material available to us at the present does not justify the conclusion that inherited genetic differences are a major factor in producing the differences between the cultures and cultural achievements of different peoples or groups. It does indicate, on the contrary, that a major factor in explaining such differences is the cultural experience which each group has undergone.

7. There is no evidence for the existence of so-called "pure" races. Skeletal remains provide the basis of our limited knowledge about earlier races. In regard to race mixture, the evidence points to the fact that human hybridization has been going on for an indefinite but considerable time. Indeed, one of the processes of race formation and race extinction or absorption is by means of hybridization between races. As there is no reliable evidence that disadvantageous effects are produced thereby, no biological justification exists for prohibiting intermarriage between persons of different races.

8. We now have to consider the bearing of these statements on the problem of human equality. We wish to emphasize that equality of opportunity and equality in law in no way depend upon the assertion that human beings are in fact equal in endowment.

9. We have thought it worth while to set out in a formal manner what is at present scientifically established concerning individual and group differences.

    1. In matters of race, the only characteristics which anthropologists can effectively use as a basis for classification are physical (anatomical and physiological).

    2. Available scientific knowledge provides no basis for believing that the groups of mankind differ in their innate capacity for intellectual and emotional development.

3. Some biological differences between human beings within a single race may be as great or greater than the same biological differences between races.

4. Vast social changes have occurred that have not been connected in any way with changes in racial type. Historical and sociological studies thus support the view that genetic differences are of little significance in determining the social and cultural differences between different groups of men.

5. There is no evidence that race mixture produces disadvantageous results from the biological point of view. The social results of race mixture whether for good or ill are to be traced to social factors.

The two UNESCO statements were signed by the leading scientists of the world.

In their monumental work *Fossil Men*, Marcellin Boule and Henri V. Vallois conclude: "Man, in spite of his superior attributes, takes his place in the scheme of general organization, and forms no exception among living creatures. In stepping backwards from history to prehistory and from prehistory to geological times, we see that always he is subject to the laws which regulate the evolution of every living thing. And it is because of this that human paleontology and general paleontology are alone found to be capable of guiding us to a full understanding of the true place of Man in nature. His true superiority of a purely intellectual order, gradually acquired in the course of a slow and laborious evolution, now enables him to raise a corner of the veil which conceals from him both the lowliness of his origin and the glory of his ascent."

REFERENCES

Boule, Marcellin, and Henri Vallois. *Fossil Men*. New York: Dryden Press, 1957.

Buxton, L. H. D. *The Peoples of Asia*. New York: Knopf, 1925.

Clark, Wilfrid E. Le Gros. *Man-Apes or Ape-Men? The Story of Discoveries in Africa*. New York: Holt, 1967.

Coon, Carleton S. *The Living Races of Man*. New York: Knopf, 1965.

Coon, Carleton S. *The Origin of Races*. New York: Knopf, 1973.

Dixon, Roland B. *The Racial History of Man*. New York: Penguin, 1925.

Dunn, Leslie C. *Heredity, Race and Society*. New York: Penguin, 1952.

Firth, Raymond W. *Human Types*. New York: New American Library, 1957.

Gann, Stanley M. *Human Races*. Springfield, Ill.: Thomas, 1971.

Goldsby, Richard A. *Race and Races*. New York: Macmillan, 1971.

Hadden, Alfred C. *The Races of Man and Their Distribution*. New York: Stokes, 1925.

Howells, William W. *Mankind in the Making*. Garden City, N.Y.: Doubleday, 1959.

Isaac, Glynn, and Elizabeth R. McConn. *Human Origins: Louis Leakey and the East Africa Experience*. Menlo Park, Calif.: Benjamin, 1976.

King, James C. *The Biology of Race*. New York: Harcourt, 1971.

Klineberg, Otto. *Race Differences*. New York: Harper, 1935.

Koenigswald, G. H. B. von. *The Evolution of Man*. Ann Arbor, Mich.: University of Michigan Press, 1976.

Leakey, Richard E., and Rogers Lewin. *Origins*. New York: Dutton, 1977.

Montagu, Ashley. *Man's Most Dangerous Myth: The Fallacy of Race*. New York: Oxford University Press, 1974.

Montagu, Ashley. *Race, Science and Humanity*. New York: Nostrand, 1963.

Montagu, Ashley. *Statement on Race*. New York: Oxford University Press, 1972. (Includes UNESCO statements on race.)

CHAPTER

7

Atlantis—Lost Continent

In *The Secret of Atlantis* Otto Muck asserts that "more books have been written about Atlantis than about any subject short of the Bible, some twenty thousand." More modest estimates have placed the figure at about five thousand. In any case, there is no question that the tradition of a lost continent has intrigued a host of writers since ancient times. A bibliography published in 1926 lists 1,700 works.

The sole source of the Atlantis story is two dialogues by Plato—poet, philosopher, statesman, and teacher of ancient Athens. In *Timaeus*, one of his greatest works, Plato is concerned with the creation of the universe and the nature of man according to Pythagorean philosophy. The second dialogue, *Kritias,* was derived from an account that Plato had heard while traveling in Egypt. The story told of a mighty empire called Atlantis, centering upon an island west of the Pillars of Herakles (the Strait of Gibraltar), larger than North Africa and Asia Minor combined, and surrounded by smaller islands, the total some 800,000 square miles in area.

The kings of Atlantis were prolific and the land rich in vegetation, minerals, and elephants. Consequently, Atlantis in time became a mighty power, with the capital city of Atlantis on the south coast of the continent. A coral of enormous size ran straight through the city and a great rectangular irrigated plain, 230 by 340 miles in dimension, was at the north end. On the central island, there were erected a royal palace, a temple, parks, barracks, racecourses, and other public facilities.

For many centuries the Atlanteans were virtuous, but then they began to suffer a moral decline. No longer satisfied with ruling their own vast territory, they tried to conquer the whole Mediterranean region. Zeus, the king of the gods, observed their wicked ambitions and greed and decided to punish them. He sent a great earthquake and flood which devastated the whole area and caused Atlantis to sink beneath the waters of the Atlantic Ocean.

To skeptics, Plato asserts that the story of Atlantis is true historical fact. At the start of *Kritias*, he says, "Listen then to a tale which, though passing strange, is yet wholly true, as Solon, the wisest of the Seven, once upon a time declared." Further, at the end of the *Kritias* narration, Socrates says, "And the fact that it is not an invented fable but a genuine history is all-important."

A. G. Galanopoulos and Edward Bacon in their *Atlantis: The Truth Behind the Legend* note the variety of opinions many critics have expressed about the Atlantis story, grouping them in four main categories:

1. Those who accept Plato's story as perfectly true.

2. Those who believe that Atlantis did exist, but not in the Atlantic Ocean, and suggest various other localities.

3. Those who claim that the Atlantis story is a compilation of legends and historical facts relating to various peoples and different ages.

4. Those who believe that the story is fictitious and an invention by Plato.

A discrepancy has been pointed out in Plato's chronology for Atlantis. According to his account, Atlantis predates his own time by 9,000 years. The Athens that Plato describes could not have existed at that early date. A well-governed, well-constructed, and well-protected state could not have existed until several thousand years later. Athens was a prototype for Atlantis. Immanuel Velikovsky, in his *Worlds in Collision*, suggests that the most probable date of the sinking of Atlantis would be 900, not 9,000 years before Solon, an Athenian statesman (638-558 B.C.).

A favorite theory among students of Atlantis is that at one time there existed an Atlantic continent or land bridge between the Americas and Europe. Such a land connection would explain the bones of mammoths or elephants, lions, tigers, camels, and primitive horses having been found in America, even though none of these were here when Europeans arrived. Looked at from a geological point of view, during the Paleozoic era, 200 million years ago, the continents were grouped together into one or two super-continents. The super-continents drifted apart in different directions to become the present continents. By the end of the Mesozoic era, seventy million years ago, the continents had come close to their present positions.

If it is assumed that a great continent once stood in the Atlantic Ocean between Europe and America, and was then submerged, it is theorized that there are survivals. The nine islands that make up the Azores would then be the peaks of the highest mountains on what was once the island of Atlantis. Other remains would include Puerto Rico and the Bermuda, Canary, and Cape Verde islands.

Throughout geological history, there is convincing evidence that the surface of the earth has been the scene of emergences and submergences of land. Thus, the Sahara Desert was once a sea, and the Mediterranean, with its underwater peaks and valleys, was once land. Fossils of sharks have been found in the Rocky Mountains, fish fossils in the Alps, and oyster fossils in the Allegheny Mountains.

Ever since Plato, speculation has prevailed whether the accounts contained in *Timaeus* and *Kritias* were legendary, allegorical, or variants of the flood stories found in the mythologies of pagan races all over the world as well as in the Bible. Was there really a great civilization in ancient times on a vast land mass now submerged in the Atlantic Ocean? Answers have been sought in mythology, comparative religion, geography, geology, and archeology.

Skeptics about the existence of Atlantis, all the way back to Aristotle, have expressed strong reservations about Plato's missing continent. As Charles Berlitz states in *The Mystery of Atlantis*, "The established academic historical community, and to a somewhat lesser degree the scientific world, has long regarded the question of Atlantis with skepticism, disbelief, and even a certain amount of hilarity ... the established world of research and history remains unconvinced for lack of more concrete proof." Martin Gardner, citing the consensus of geologists' opinions, notes, "There is unanimous agreement that no great continental sinkings have occurred within the relatively brief time man has been on the scene. There is, in fact, not a shred of reliable evidence, geological or archeological, to support the myth of Atlantis." The views of two geologists are of special interest. Dorothy B. Vitaliano maintains, after examining all the evidence, "From the geological point of view, I fear that Atlantis must be considered just another of the myths of Plato." Herbert E. Wright, Jr., adds, "In view of the numerous difficulties in relating glacial events to short-term global sea-land changes, Atlantists will have to look elsewhere for their catastrophes." L. Sprague de Camp, in his *Lost Continent*, suggests, "So let us, once and for all, grant decent burial to all the theories of Atlantis that a great island did, as Plato says, sink beneath the sea in a day and a night of storm and earthquake."

If the concept of Atlantis lacks reality, what was Plato's intent in writing it? One possibility is that Plato was writing a political, historical, and scientific romance, a kind of science fiction story. The theme was in accord with the views of geography, geology, and history prevailing in Plato's time.

It has been pointed out, too, that Atlantis was identical in most respects to the ideal state which Plato described in the *Republic*. In *Timaeus* and *Kritias*, Plato was putting some of his *Republic* theorizing on a more practical level by giving an example of how it might work in practice.

How can one explain the fascination that the myth of a lost continent has had over the minds of so many scholars and writers? The literature continues to grow, despite debunking efforts. There are, of course, psychological factors, one of which is an element of escape – dreaming about a vast, mythical land of wonder, a happy perfection once possessed by mankind. Martin Gardner concludes, "It is safe to say that speculations about Atlantis will occupy the minds of pseudo-archeologists for many decades to come."

REFERENCES

Berlitz, Charles. *Atlantis, the Lost Continent*. New York: Fawcett, 1984.

Berlitz, Charles. *The Mystery of Atlantis*. New York: Avon, 1975.

Braymer, Marjorie. *Atlantis*. New York: Atheneum, 1983.

De Camp, L. Sprague. *Lost Continent*. New York: Dover, 1970.

Galanopoulos, A. G. *Atlantis*. Indianapolis, Ind.: Bobbs-Merrill, 1969.

Luce, John V. *Lost Atlantis*. New York: McGraw-Hill, 1969.

Muck, Otto. *The Secret of Atlantis*. New York: New York Times Books, 1978.

Ramage, Edwin. *Atlantis, Fact or Fiction?* Bloomington, Ind.: Indiana University Press, 1978.

Spence, Lewis. *The History of Atlantis*. Philadelphia: University Books, 1965.

CHAPTER

8

Noah's Ark

The story of Noah's Ark and a great flood is a notable feature of Genesis in the Old Testament. It takes as much space in Genesis as the days of creation, Adam and Eve, and the Garden of Eden. According to the biblical account, mankind had become so wicked that God determined to destroy living things. An exception was made of Noah and his family because in the eyes of the Lord he was "a just man and perfect." Noah was commanded to build an ark and to take into it his own family and a pair, male and female, of every variety of living creature. The dimensions of the ark were specified: 300 cubits long (in English measure, a cubit equals eighteen inches), fifty cubits wide, and thirty cubits high. There were to be three stories, and the ark would be constructed with a door and window.

Having done as God commanded, Noah, along with his family and the animals, were safely on board when the rains came. It continued to rain for forty days and forty nights, during the course of which every living substance was destroyed. The water prevailed upon the earth for 150 days and then began to recede. First a raven and then a dove were released by Noah to determine whether the earth was dry. The ark landed on Mount Ararat, and its passengers, human and animal, went forth to repopulate the earth.

In *The Lost Ship of Noah*, Charles Berlitz notes that "the story of the Great Flood and the survival of man, together with his family and animals in an ark or ship has existed in legends and folktales for more than 3,000 years and perhaps as far back as 11,000 years."

Ancient Babylonia had two deluge stories, one traced to about 2000 B.C. and containing many parallels to the Genesis account. Some scholars believe that both the Babylonian and biblical stories are derived from a common Semitic source. Deluge stories are also found in Egypt, India, Persia, Syria, and Greece, and among such widely separated peoples as the American Indians and the Polynesians. Variations of the legend among ancient nations and tribes number more than 600.

Many scholars have inclined toward skepticism in viewing the Genesis story. If the legend has validity, especially in the light of its world-wide distribution, what could have been the cause of such a catastrophe? Speculations have ranged from a world-wide deluge or earthquakes to such far-fetched theories as collision with an asteroid, slipping and melting of glaciers and ice caps, shifting of mountain ranges, and sinking of land masses into the ocean.

An actual fact, in any case, is the existence of Mount Ararat in eastern Turkey, the highest, most formidable, and most difficult to climb of Armenian mountains. The giant 12,000-foot peak, of volcanic origin, dominates the barren semiarid plains of eastern Turkey.

The fate of Noah's Ark has intrigued historians, explorers, archeologists, religious leaders, and others for generations. For the past two centuries, teams of mountaineers have climbed Ararat in search of the ark. Travelers on horseback, mules, and camels have claimed to have observed a great ship near the top of the mountain. As early as 1916, members of the Russian Imperial Air Force flew over the mountain and spotted what they took to be the half-submerged hull of a ship in a half-frozen lake. According to a magazine article written by Lieutenant Roskovitsky, the plane's pilot, based on reports of engineers ordered by the tsar to investigate, "The Ark was found to contain hundreds of small rooms, and some rooms that were very large, with high ceilings ... other rooms were lined with tiers of cages." Doubts have been cast on the truth of the Russian's story.

Other expeditions, from 1800 on, have failed to find the wreck of the ark. There is belief now that the ark has slipped down the mountain and may have fallen into a deep crevasse, frozen into one of Ararat's glaciers. In any case, after intensive research at the site, Berlitz "is convinced that definite proof of the historical validity of the Flood and the Ark now exists."

One matter requiring proof is how a ship could survive for thousands of years on top of a mountain, exposed to earthquakes, glaciers, and freezing temperatures. Great cities and ancient civilizations built of wood, iron, copper, brick, and all other building materials except stone have vanished with time. How could a wooden ship such as the ark have survived through the centuries? For the ark, the accepted explanation is freezing. The glaciers on Mount Ararat are cold enough to preserve a craft constructed of heavy timbers and caulked "within and without with pitch," as stated in Genesis. Buried under the ice, the ark is impervious to decay, like such extinct animals as the mammoths of Siberia and saber-toothed tiger, belonging to the Tertiary period of geologic time.

The story of Mount Ararat and the missing ark has become a focal point of discussion between religious conservatives and modernists, or between creationists and evolutionists. If the existence of the ark can be completely verified, it would go far toward authenticating the biblical account and convincing the skeptics. As Berlitz comments, "It could modify our concept of ancient history, reassess the geological history of the world, and reaffirm the traditional religious beliefs of a great section of the world's people."

Meanwhile, the search goes on, the great prize seemingly almost within reach.

Skeptical biologists have pointed out the massive problems that must have confronted Noah in trying to obey the mandate that he had received. Noah would have had to find room for about a million species of insects, twenty-five thousand species of birds, twenty-five hundred species of amphibians, six thousand species of reptiles, and several thousand species of mammals. All had to fit inside an ark with only 43,800 cubic meters of space, together with their food of diverse nature.

In Search of Noah's Ark, a highly popular film produced in 1977, tries to answer the question of whether the ark actually exists on top of Mount Ararat. "The movie is in part dramatization," *Newsweek* reported, "and in part interviews with a parade of scientists, historians, and fundamentalist preachers." The conclusion: "Most Biblical scholars take neither the film nor the search for the Ark seriously."

REFERENCES

Allen, Don C. *The Legend of Noah.* Urbana, Ill.: University of Illinois Press, 1963.

Berlitz, Charles. *The Lost Ship of Noah, in Search of the Ark at Ararat.* New York: Putnam, 1987.

Filby, Frederick A. *The Flood Reconsidered.* London: Pickering, 1971.

LaHaye, Tim, and John D. Morris. *The Ark on Ararat.* Nashville, Tenn.: Nelson, 1976.

Montgomery, John W. *The Quest for Noah's Ark.* Minneapolis, Minn.: Bethany Fellowship, 1972.

Navarra, Fernand. *Noah's Ark Expedition.* Plainfield, N.J.: Logos Inst., 1974.

Parrot, Andre. *The Flood and Noah's Ark.* London: Coverdale House, 1955.

CHAPTER

9

Great Wall of China

American astronauts, viewing the earth from outer space, reported that the only man-made structure they could identify back on earth was the Great Wall of China. In view of its enormous dimensions, that fact is easily understandable.

The Great Wall is a massive fortification running along the northern and northwestern frontier of the country. It was erected by the Emperor Chin Shih Huang Ti shortly after he had reunited different sections of the country into an empire in 228 B.C.

The actual task of constructing the greatest engineering achievement of the ancient world was assigned to an outstanding military officer, Meng T'ien. The wall was completed in seven years, about 214 B.C. According to one source, 300,000 laborers were available for the gigantic task. Other sources indicate that more than a million men may have worked on the wall during the years of construction. Meng T'ien had the advantage of being able to incorporate long sections of earlier walls into it.

The physical conditions under which the wall was built were extremely difficult. The climate varied from broiling heat to winter snow. The wall ascended mountains and crossed endless plains, undeterred by freezing gales, blinding sandstorms, or raging storms. The Great Wall comprises three main sections. It begins at Shanhaikuan on the Yellow Sea, passing through mountainous terrain, where it is built of brick and stone. The second section is the eastern rampart, now largely destroyed, running from the Yellow River's eastern arm to the western end of the great river's loop.

This section was built of loess, a poor material from which to build a wall that is subject to destruction by the elements. From there, the route skirts the Yellow Sea to Liangchow, the desert loop. The third section runs through mountainous country from Liangchow to the western terminal point to Kia-Yu-Kuan.

The distance covered by the three sections is 1,850 miles, but the actual length of the wall is 3,930 miles, including a three-hundred-mile extension west of Kia-Yu-Kuan, a four-hundred-mile loop through Manchuria, and more than a thousand miles of inner walls.

The usual explanation for construction of the Great Wall is that it was designed to keep out barbarians from the north. It appears, however, that the emperor who ordered construction of the wall was as concerned with keeping his own people in as keeping nomads out. The wall also absorbed the energy of a restless army of landless men. Public works had long been a means by which Chinese kings had maintained order, and here was an incredibly large public works project.

The dimensions of the Great Wall are staggering. For instance, the wall contains enough material to build a barrier eight feet high and three feet thick that would completely encircle the globe at the equator. A member of the British embassy in 1790 estimated that the Great Wall contained more brick and stone than all the buildings of the United Kingdom.

The route of the Great Wall took it through soft loess, a loamy deposit formed by the wind, and along the rims of mountains thousands of feet high. At one point, the wall was eighty feet below sea level, at another, 10,000 feet above.

The building of walls is an ancient tradition in China. Walls surround every Chinese city and are a basic feature of Chinese communities. In northern China, there is scarcely a village of any size or age without at least a mud wall. The walls are generally kept in good condition, better cared for than other buildings. The Chinese word ch'eng can mean either a city or a wall.

In *The Great Wall of China* (London, 1909), William Edgar Geil asserts, "The Wall is not for modern use. It is an ancient fossil—the largest fossil on the planet." The futility of the wall's primary purpose—to turn back invaders—is shown by the fact that barbarians ruled China for 800 out of the last 1,500 years.

An anonymous writer once declared that the Great Wall was "the greatest example of wasted military effort in the history of the world." Another commentator, Robert Silverberg, in *The Great Wall of China*, disagrees, pointing out that "the Wall has always been more than just a mighty rampart; it has been a symbol of China, the concrete and tangible representation of the unity of that vast land.... The story of China is the story of the magnificent Long Rampart that gave her identity."

Inevitably, there have been comparisons between the building of the Great Wall of China and the Great Pyramid of Egypt. The latter is reputed to have kept thousands of men busy for twenty years. Construction of the Great Pyramid required an immense quantity of material, as well as time and manpower. The physical problems encountered in erecting the Great Wall, however, almost certainly exceeded in difficulty those which had to be surmounted in building the Great Pyramid.

A construction project as monumental as the Great Wall naturally stirred the Chinese imagination, and a considerable body of folklore grew up around it. Over the past twenty-two centuries, many legends have come down concerning the building of the wall. Among the tales are explanations of why there are occasional deviations from a straight line in the route followed. Many of the legends of the Great Wall are love stories. One ancient story tells of thousands of workmen who died during the construction and were buried in and under the Wall.

An unexplained mystery is why the famous traveler Marco Polo never once mentioned the Great Wall in his account of extensive travels in China in the fourteenth century. He could hardly have failed to see the wall, but *The Travels of Marco Polo* makes no reference to it.

REFERENCES

Dalin, Chong. *The Great Wall of China.* Hong Kong: n.p., 1984.

Geil, William Edgar. *The Great Wall of China.* London: John Murray, 1909.

Kafka, Franz. *Great Wall of China.* New York: Schocken, 1946.

Silverberg, Robert. *The Great Wall of China.* Philadelphia: Chilton, 1965.

Warwick, Adam. "A Thousand Miles Along the Great Wall of China." *National Geographic Magazine* 43, no. 2 (February 1923).

CHAPTER

10

Pyramids, Egyptian and Mexican

PYRAMIDS IN EGYPT

When Antipater, regent of Macedonia in the second century B.C., set out to list seven wonders of the ancient world that were outstanding for their size or beauty, he ranked first the pyramids of Egypt. Of the seven, the pyramids are the only structures that are still standing.

Some ten miles west of Cairo is a rocky plateau at the end of an avenue lined with acacia, tamarind, and eucalyptus trees. The man-leveled plateau, a mile square, dominates the Nile Valley from a height of 130 feet. On the plateau, called Giza by the Arabs, stands the Great Pyramid of Cheops. To the west is the vast Libyan desert.

According to one commentator, "the pyramids of Egypt are immensely large, immensely ancient, and, by general consensus, extremely useless." Whether useful or not, the ancient Egyptians spent a tremendous amount of labor in their construction.

The Great Pyramid of Cheops, largest of them all, occupies thirteen acres, equivalent to seven city blocks in a modern city. Its construction required 2.5 million blocks of limestone and granite. The blocks weighed from two to seventy tons each, rising in 201 stepped tiers to the height of a modern, forty-story skyscraper. Using another comparison, the Great Pyramid contains more stone than all the cathedrals, churches, and chapels in England since the beginning of the Christian era.

The famous Greek traveler Herodotus visited and described the Great Pyramid in the fifth century B.C. According to his account, the pyramid required the labor of 100,000 men for thirty years. Historians are inclined to think that the number was exaggerated, since slave labor was limited during that period.

The Great Pyramid was built of limestone blocks quarried near the site and faced with a finer limestone from across the Nile. The granite for passages and chambers was brought down the river from Aswan. The Egyptians raised and turned stone blocks by use of a lever. Transportation of the blocks was on sledges, probably on rollers laid in front of them, and then on plank skidways pulled up the pyramid face.

There have been many attempts to explain the function or purpose of pyramids; possibly, they were astronomical observatories, grain storerooms, refuges from the Great Flood, repositories of divinely inspired prophecies, or even the work of interplanetary visitors. There is solid evidence, however, that the pyramids served as funerary monuments for the early pharaohs. The ancient Egyptians believed in an afterlife during which the soul returned to the body. For that reason, the body was carefully preserved through mummification. Mummies of royalty and nobility were placed in tombs with objects for use in the afterlife. A large body of mortuary priests was maintained in connection with the pyramids.

While it is recognized that the principal pyramids served as royal mausolea, it seems certain that they had other uses. It has been demonstrated that sightings from inside the Great Pyramid could be used to chart the sky. Some astronomers believe that the structure was intended to be a mathematical model of the Northern Hemisphere. Calculations have revealed a relationship of the pyramid's dimensions to various dimensions of the earth.

A nineteenth-century mathematician, Piazzi Smyth, theorized that the Great Pyramid had been designed as a huge sundial whose shadows could indicate the seasons and the length of the year. By his reckoning, the pyramid had been intentionally located, oriented, and sloped to reveal these facts. A French astronomer, Jean Baptiste Biot, came to a similar conclusion, stating that the Great Pyramid "functioned as an immense sundial which marked annually the periods of the equinoxes with an error less than one day, and those of the solstices with an error less than a day and three quarters."

It is estimated that Egyptian kings built more than eighty pyramids. Most were constructed over a period of five centuries along the west side of the Nile, within fifty miles of Giza. To guard against looters, tombs were made increasingly large and strong, until the pyramid became the standard royal tomb. Its passages were blocked and its entrance covered to make it safe from tomb robbers. Despite such precautions, nearly all Egyptian pyramids were looted in ancient times. In addition, the building blocks were often taken for other building projects.

Related to the mysteries surrounding the pyramids is the Great Sphinx, which is located about twelve hundred feet southeast of the Pyramid of Cheops. The Sphinx is carved from a single sandstone knoll. Its dimensions are 240 feet in length, 66 feet in height, and 13 feet 8 inches in width. The figure is half lion, half human, symbolizing the junction of the constellations Leo and Virgo.

An international sensation was caused in 1922 by the discovery of the tomb of Tutankhamen in the Valley of the Kings. A pharoah of the eighteenth Dynasty of Egypt, Tutankhamen reigned from about 1361 to 1352 B.C. Tutankhamen's tomb was the richest Egyptian tomb ever found and has been called one of the most spectacular archeological finds of the twentieth century. It had not been looted and was almost intact when discovered by a British archeologist, Howard Carter. Among 2,000 objects found in four small chambers were a gilded throne inlaid with silver and semiprecious stones, gilded beds and chariots, chests of jewelry, alabaster vases, weapons, and religious articles. Tutankhamen's body lay inside a solid gold coffin. After a world tour, the contents were placed in the Egyptian Museum at Cairo. The tombs of more than forty pharaohs are carved out of the hills and cliffs surrounding the Valley of the Kings, near the site of the ancient city of Thebes.

PYRAMIDS IN AMERICA

The Maya Indians of the Yucatan region of Mexico were builders of pyramids in Middle America from the fourth through ninth centuries. Their pyramid style was adopted later by the Toltecs, Aztecs, and Incas. American pyramids were formed of earth and adobe and faced with stone slabs, with great stairways leading to the top. Flattened areas on top served as bases for shrines, not tombs. A famous example, the Mayan Pyramid at Uxmal, had a base 240 feet by 180 feet and was eighty feet high. One of the largest pyramids, the Pyramid of the Sun at Teotihuacán, Mexico, has a base more than 700 feet square and is 216 feet high. Another huge pyramid, built over a period of hundreds of years at Cholula, Mexico, had a base that covered forty-two acres.

Pyramids and the human sacrifices on top of them were discovered by the conquistadors to be standard features of life throughout Mexico. With the sacrifice of thousands of victims each year, cannibalism was practiced quite generally. The Conquest of Mexico brought human sacrifice to an abrupt end. It is not known how many sacrifices occurred earlier, but native records show that at the dedication of the great temple of Tenochtitlán in 1487, 20,000 victims were slain. Elsewhere, thousands of skulls were found in towns and villages.

Why did the natives not revolt against this hideous custom? Reasons for submission included religious fanaticism and the Aztec belief in life after death. The victims believed they would proceed directly to heaven, the abode of the gods. There were protests when Cortez forbade the practice.

Theories have been suggested that the American pyramids were inspired by those in Egypt through some connection in the remote past. No solid evidence exists, however, of any association. The Aztec and Mayan pyramids can be compared in size with the later Egyptian ones, but they are far inferior to the huge monuments from the height of the Egyptian pyramid age.

REFERENCES

Edwards, I. E. *The Pyramids of Egypt.* New York: Viking Press, 1972.

Fakhry, Ahmed. *The Pyramids.* Chicago: University of Chicago Press, 1961.

Grinnell, Leslie V. *Egyptian Pyramids.* Gloucester, England: Bellows, 1947.

Kubler, George. *The Art and Architecture of Ancient America.* New York: Penguin, 1976.

Leclant, Jean. *In the Steps of the Pharaohs.* New York: Hasting House, 1959.

Mendelssohn, Kurt. *The Riddle of the Pyramids.* New York: Praeger, 1924.

Stewart, Desmond, and others. *The Pyramids and Sphinx.* New York: Newsweek, 1971.

Tompkins, Peter. *Secrets of the Great Pyramid.* New York: Harper, 1971.

CHAPTER

11

Stonehenge

It would not be difficult to defend the proposition that Stonehenge is by far the most impressive prehistoric monument north of the Egyptian pyramids. This celebrated landmark, consisting of huge stones and boulders, is located on Salisbury Plain in Wiltshire County, England, about eight miles north of Salisbury, eighty miles west of London, and only thirty miles north of the English Channel.

Stonehenge is believed to date back to at least 2000 B.C. The original builders are unknown. The outer circle, about 100 feet in diameter, was probably originally composed of thirty huge stones, each one fourteen feet high, seven feet wide, and 3½ feet thick. Every two stones support a ten-foot-long stone slab. The next inner circle, seventy-five feet in diameter, was made of stones of much smaller size. Within this circle, five groups of huge stones were arranged in a horseshoe pattern. A similar pattern of smaller stones forms the last group, which surrounds a flat altar stone, the center of the entire monument. Surrounding the monument is a circular ditch and bank, 300 feet in diameter. Two other stones, the large flat slaughter stone inside the entrance and the upright Friar's Heel or Hele stone, complete the structure of Stonehenge.

Two types of stone were used in the construction of Stonehenge. One type was probably obtained from the Marlborough Downs, about twenty miles away, and the other, in the form of 270-ton blocks, from a quarry in the Prescelly Mountains of Wales more than 100 miles away.

An unsolved question, about which there are various theories, is how the huge bluestones from southern Wales were transported to Salisbury Plain. The route most widely accepted is mainly by sea. Going by sea from Milford Haven to the mouth of the Avon and up the Avon to Amesbury brought the bluestones to within two miles of Stonehenge. In any case, transporting 300 tons of stone from Pembrokeshire on the Irish Sea and the Bristol Channel to Stonehenge was an extraordinary feat—one of the most striking facts about Stonehenge. The effort, knowledge, skill, and courage required were probably inspired by a religious motive. Why were the stones brought from so far away? One belief was that the bluestones had curative powers, or that they were thought to possess a special holiness.

The so-called sarsen stones used in the horseshoe inner circle presented a different problem. These stones had to be lifted out of the ground, twenty miles away, and placed on rollers, then probably loaded on sledges and dragged across the ground with ropes. Weighing thirty tons each, it has been estimated that at least fifteen hundred men working five years would have been required to transport the sarsen stones to Stonehenge.

Why was Stonehenge built? There has been much speculation about the purpose. It has been called many things, including a Roman temple, a tomb, a Druid temple, and a center for sun worshippers. Most archeologists consider it to have been a sun temple, built by an agricultural people.

Supporting the idea of Stonehenge as a solar temple, it has been noted that for a person standing at the center of Stonehenge on the summer solstice, about June 21, the sun rises from behind the heel stone. Certain pairs of stones were aligned with the rising and setting positions of the sun on the first days of fall, winter, and spring. The priests, who performed various rituals and sacrifices, by determining the position of the sun could determine the seasons and thus declare the most suitable time for planting. Outside the sarsen circle is a ring of fifty-six pits, called Aubrey holes, a primitive device for predicting lunar eclipses, named after the discoverer.

In *The Mysteries of Stonehenge*, Fernand Niel offers convincing support for the idea that the unique character of Stonehenge shows that it is the work of a single man. The people who built the monument supplied the labor, but would have been incapable of conceiving the design for Stonehenge. According to Niel, the brilliant architect most likely came from one of the countries around the eastern part of the Mediterranean. "Only there could one have learned enough geometry, architecture, and astronomy to design such a monument."

Archeologists believe that Stonehenge has gone through several stages. In the first phase it was used by people of the late Stone Age, probably for religious gatherings. About 1700 B.C., a people with more advanced technical knowledge brought the forty-ton bluestones from Wales and erected the first stone structure. Several centuries later an early Bronze Age people rebuilt the monument by transporting the sarsen stones, weighing up to fifty tons each, overland 200 or more miles. The last builders may have had some contact with the advanced civilization of Mycenaean Greece. Tradition associates Stonehenge with the Druids, who may have used it for ceremonial purposes but had no part in its construction.

Nowhere else in the world is there anything comparable to Stonehenge. This fact has doubtless accounted for the voluminous literature relating to it—more than six hundred books and articles by the year 1900 and double that figure in the years since. One of the first expert commentators was the famous sixteenth-century English architect Inigo Jones, who became fascinated by Stonehenge and made a comprehensive study of it, illustrated with admirable drawings and an attempt to depict its original state. The first writer known to have mentioned Stonehenge was Henry of Huntington, author of *Historia Anglorum*, written in the eleventh century. Describing the wonders of Great Britain, he says, "The second one is at Stonehenge, where stones of amazing size are erected like porches, so that one door seems to have been placed on another. It would be difficult to say by what means they were raised so high and why they were put in that place." Stonehenge has continued to mystify countless observers ever since.

In 1963 American astronomer Gerald Hawkins described Stonehenge as a sort of prehistoric computer which could make intricate calculations of sunrises and sunsets, the movements of the moon, and solar and lunar eclipses. Hawkins added that by moving a marker stone around an inner circle of holes once a year, priests or astrologers could calculate suitable times for crop planting, predict weather cycles, and practice divination.

UPDATE

Cullen Murphy, writing in the *Atlantic* (259: [June 1987]: 14-16), states recent findings have concluded that Stonehenge is 5,000 years old, based on radiocarbon dating, making it one of the most ancient of early architectural monuments.

REFERENCES

Atkinson, Richard J. C. *Stonehenge.* New York: Hamilton, 1956.

Chippindale, Christopher. *Stonehenge.* Ithaca, N.Y.: Cornell University Press, 1983.

Crampton, Frederick. *Stonehenge of the Kings.* New York: Day, 1968.

Hawkins, Gerald S., and John B. White. *Stonehenge Decoded.* Garden City, N.Y.: Doubleday, 1965.

Niel, Fernand. *The Mysteries of Stonehenge.* New York: Avon, 1974.

Stone, E. H. *The Stones of Stonehenge.* London: Hamilton, 1924.

Stover, Leone, and Bruce Kraig. *Stonehenge: The Indo-European Heritage.* Chicago: Nelson-Hall, 1978.

CHAPTER

12

Easter Island

One of the most remote and inaccessible bits of real estate on the earth's surface is Easter Island, a pinpoint in the South Pacific. Nevertheless, it has long held a special fascination for anthropologists and ethnologists.

The discovery of Easter Island was first reported by a Dutchman, Jacob Roggeveen, commodore of a squadron of three vessels in 1722. The island was christened "Easter" in honor of the holy day that preceded the discovery. The day also marked the beginning of Easter Island's tragic history. Many of the natives came out to visit the Dutch ships. When some of them were caught thieving, the crew members fired on them and many fell. The arrival of "civilization" was accompanied by bloodshed.

Later in the eighteenth century, ships from France, Peru, England, Russia, and America visited the island. Noteworthy was Captain James Cook, who explored the land in 1774 and left a detailed account of its geographical situation, the inhabitants' manners, customs, government, and religion, and a description of the gigantic statues found on the island.

The most disastrous contact the Easter Islanders had with foreign intruders occurred in December 1862, when a fleet of six Peruvian ships appeared with the purpose of kidnapping the men and carrying them off to slavery in the flourishing guano quarries off the coast of Peru. About one thousand of the wretched slaves were imprisoned on the Chinchos Islands. There were protests by the French and English governments. By the time the Peruvian government ordered the release of the prisoners, however, 80 percent of them were dead from ill treatment, undernourishment, or disease. About one hundred were taken back to Easter Island and nearly all died of smallpox during the voyage. The fifteen who survived brought the dreaded disease with them and more than half the population of the island perished within a few months. Of the five thousand people who had lived on the island, no more than about six hundred were left.

A few geographical facts: Easter Island is 2,700 miles from Tahiti and 2,600 miles from Valparaiso. The Galapagos archipelago lies some 2,000 miles to the northeast. To the south there is the Antarctic. The area of the island is 45.5 square miles. The triangle-shaped island is volcanic and has no harbor. Easter Island was annexed by Chile in 1888, and a Chilean company uses a high percentage of the land for grazing sheep and cattle.

The most striking monuments on Easter Island are burial platforms, called *aku*, and the statues on top of them. The *aku* was used for the exposure of dead bodies, and contained vaults for later burial of the bones. The giant statues, usually ranging in height from twelve to twenty feet and weighing up to fifty tons each, were cut from compressed volcanic ash, a soft stone. The quarry for the statues was located inside and outside the large crater of a volcano called Rano-Raraku. Representing the upper half of a body with an upturned face and long ears, each figure was topped with a tall cylindrical hat or crown cut from red volcanic tuff that had been quarried from a different crater. It is believed that the statues were erected in honor of ancestors. They stood in a single row facing inland on a wall that projected toward the sea. Most of them were standing in the eighteenth century, as reported by early observers, but later they were thrown down and many were broken.

An unexplained mystery is why more than eighty statues were left unfinished. Everything seems to have stopped suddenly, in a single day, as though hit by some enormous disaster. There are various conjectures to explain the sudden stop: a fratricidal war, a natural phenomenon, such as a meteor that fell too close, or perhaps a terrible disease.

Also highly mysterious is the question of how the statues were moved to the *aku*, some of them twelve or fifteen miles from the quarry. The biggest statues weighed as much as fifty tons and stood as much as seventy-two feet high. Furthermore, the terrain was stony and uneven, and it would appear that the statues would have been damaged in moving; actually, they were not even scratched. One theory is that transportation was on sledges made of banana tree stems. A native account states that the statues were dragged into position with ropes made of native hemp hibiscus fiber, with round pebbles placed underneath to serve as rollers. Seaweed may also have been used to minimize friction with the ground.

Habitations on Easter Island were generally constructed of perishable materials. Natural caves were also widely used. A sacred village, called Orongo, contains forty-eight houses built entirely of stone. The village was associated with a bird cult, which played an important part in native life. There was an annual competition to procure the first egg laid on an outlying islet by the migratory sooty tern. Numerous rock carvings in the vicinity depicted a bird-headed man, a symbol that wandering sea birds were an important source of food on the island. Care was taken, however, not to molest the sooty terns in their nesting. Otherwise, the birds would have been exterminated.

An enigma to which no satisfactory answer has yet been found is the existence on Easter Island of a form of writing on wooden tablets, ideographic or pictographic, that no one has been able to decipher. Virtually all the learned men who understood this writing died in 1862, at the time of the Peruvian raid. The original culture also died out with the coming of the first missionaries. At one time, knowledge of the writing was reserved first for the royal family, second for the chief of six regions, and finally for the scholars. The script is believed to have religious significance. There are symbols for the fish, the bird, the crab, the sun, and the female emblem of fertility. An odd feature is that only alternate lines are right side up so, in reading, a tablet has to be reversed at the end of each group of characters. The exact meaning of the symbols and method of interpretation have been lost, probably beyond recovery.

In early days, Easter Islanders were free from most diseases. Because they had no resistance to the diseases of white men, smallpox came close to wiping out the native population, venereal diseases afflicted many, and a high percentage were victims of tuberculosis. Leprosy was introduced from China by way of Tahiti, and Easter Island now has a leper colony. Cannibalism has given the natives a good working knowledge of human anatomy and their methods of treating wounds received in war are surprisingly efficient.

Racially and linguistically there is no doubt that the Easter Islanders are predominantly Polynesian. The earliest settlers may have arrived in the fourteenth century or perhaps earlier. There is no evidence of a culture earlier than the present one.

REFERENCES

Brown, J. MacMillan. *The Riddle of the Pacific*. London: Unwin, 1924.

Casey, Robert J. *Easter Island*. Indianapolis, Ind.: Bobbs-Merrill, 1931.

Dos Passos, John. *Easter Island, Island of Enigmas.* Garden City, N.Y.: Doubleday, 1971.

Maziere, Francis. *Mysteries of Easter Island.* New York: Norton, 1965.

McColl, Grant Raspanvi. *Tradition and Survival on Easter Island.* Honolulu, Hawaii: University Press of Hawaii, 1981.

Métraux, Alfred. *Ethnology of Easter Island.* Honolulu: The Museum, 1940.

Routledge, Katharine S. *The Mystery of Easter Island.* London: Hazell, 1920.

CHAPTER

13

Dead Sea Scrolls

There are several versions of how a remarkable cache of ancient manuscripts was found about 1947 in desert caves on the northwest shore of the Dead Sea, near Jerusalem. The generally accepted story is that a fifteen-year-old boy of the Taamirah tribe of Bedouins, named Muhammad Adh-Dhib, was chasing a lost goat. When the goat jumped into a cave, the young boy goatherd threw a stone after it and heard the sound of breaking glass. His curiosity aroused, he crawled into the cave and found a number of scrolls wrapped in linen and deposited in jars. They were immediately labeled Dead Sea Scrolls.

Subsequently, eleven other caves in the area were searched and yielded additional manuscripts. For the most part, the scrolls are biblical and religious texts written in Hebrew and Aramaic. They are a thousand years older than the earliest previously known Hebrew text of the Bible and shed new light on Jewish religious literature and history at the time of Jesus. The scrolls are believed to have belonged to a Jewish religious community, the ascetic Essenes, a sect that existed in the Dead Sea area from about 180 B.C. to A.D. 68.

Found among the Dead Sea Scrolls were manuscripts or fragments of all the Old Testament books except Esther. In addition, Cave I, as it came to be designated, included an Isaiah scroll, Rule of the Community (a manual of Essene teachings and discipline); The War of the Sons of Light Against the Sons of Darkness, a war scroll modeled on Roman military manuals; a scroll of thanksgiving hyms; a commentary on the Book of Habakkuk; and seven other well-preserved scrolls.

From Cave III came the Copper Scroll, an apocryphal list of temple treasures and their hiding places. The main deposit of the Essene library was discovered in Cave IV, consisting of about 400 manuscripts, mostly sectarian writings. Approximately 100 are biblical, representing the Hebrew Old Testament and including a valuable scroll of the Book of Samuel. Documents from Cave XI included a large scroll with canonical, apocryphal, and unknown psalms; a copy of Leviticus in Paleo-Hebrew; and the Temple Scroll giving details for construction of the ideal Temple of Jerusalem.

Inspired by the first discoveries, archeologists began a systematic exploration of the surrounding caves. The principal finds were thousands of scroll fragments, which were taken to Jerusalem and sorted by an international team of scholars. Some fragments no larger than a fingernail were carefully cleaned and then laid out between sheets of glass on long tables. The scholars then tried to fit the pieces together like a jigsaw puzzle. Complete scrolls were softened by months of exposure to carefully controlled humidity before they could be unrolled. It has been estimated that the work of deciphering and interpreting the various writings may take as long as fifty years. Meanwhile, the Dead Sea Scrolls are housed in the Shrine of the Book in the Israel Museum. Many have not yet been published.

The Rule of the Community scroll found in Cave I is a document describing the basic constitution of the group, the qualifications for membership, the procedure for admission, the various ranks, the regulations governing the social intercourse of the brethren, and the schedule of penalties for minor and major offenses (including the extreme of expulsion from the group). There is also a sermon stating the doctrine of the two instincts of man, the good and the bad, noting the rewards and punishments appropriate to each. Finally, there are rules for military service, education, eligibility for office, and the conduct of communal meals.

A book of hymns, titled *Hymns of Thanksgiving*, also found in Cave I, has as its main themes the spiritual illumination of the elect, their privation and devotion, their inclusion in a wider communion that also embraces the angels, and the eventual triumph of God over the wicked and those who are misleading the people.

The Essenes lived a pious, ascetic, communal life, regarding themselves as people of the New Covenant. They were preoccupied with numerous baptisms, took part in a sacred common meal, and lived in constant expectation of the coming of two Messiahs and the final victory of God.

A leading biblical critic, David Noel Freedman, summed up the significance of the Dead Sea Scrolls as follows:

1. since these manuscripts are some thousand years older than previous biblical documents, an earlier form of the Old Testament may be recovered;

2. the documents provide information on Judaism at the beginning of the Christian Era;

3. the scrolls throw new light on Christian beginnings and on New Testament interpretation.... The new scrolls also support the conclusion that the New Testament represents a radical departure from Judaism.

From the scrolls it is clear that the Essenes firmly believed that they were God's elect, both as members of the chosen people and as sons of light. As a community, they believed in the teacher of righteousness as the inspired interpreter of the divine mysteries. They expected the judgment and eternal punishment of the sons of darkness, when the dominion of Belial would be brought to an end. They hoped by being cleansed of all evil to enjoy eternal felicity in the presence of God with the angelic hosts.

The Dead Sea Scrolls have been acclaimed as "the greatest manuscript discovery ever made in the field of biblical studies."

REFERENCES

Burrows, Millar. *The Dead Sea Scrolls.* New York: Viking, 1955.

Burrows, Millar. *More Light on the Dead Sea Scrolls.* New York: Viking, 1968.

Cross, Frank M., Jr. *The Ancient Library of Qumrun.* Garden City, N.Y.: Doubleday, 1961.

Davies, Arthur P. *The Meaning of the Dead Sea Scrolls.* New York: New American Library, 1956.

Gaster, Theodore H. *The Dead Sea Scrolls in English Translation.* Garden City, N.Y.: Doubleday, 1976.

Noble, Iris. *Treasure of the Caves.* New York: Macmillan, 1971.

Roth, Cecil. *The Dead Sea Scrolls.* New York: Norton, 1965.

Vermes, Geza. *The Dead Sea Scrolls.* Philadelphia: Fortress Press, 1977.

Wilson, Edmund. *Israel and the Dead Sea Scrolls.* New York: Farrar, 1978.

Yadin, Y. *The Message of the Scrolls.* New York: Simon & Schuster, 1957.

CHAPTER

14

Discovery of America

According to the best available information, the Mongoloids, who drifted into North America from Asia thousands of years ago, were the first discoverers of America and the ancestors of American Indians. These early discoverers wandered into America across the land bridge that once connected Siberia and Alaska.

Much less solidly based traditions have traced contacts between the Old World and the new to the Chinese, the Phoenicians, the Romans, the Portuguese, and others. Cyrus H. Gordon, in *Before Columbus: Links between the Old World and Ancient America*, states that "Greek classics independently and repeatedly attest transatlantic contacts between the Mediterranean and America," and adds that "sailors from the Near East reached America as early as 531 B.C." Roman contact with America around A.D. 200 is indicated by the recovery of ancient Roman coins in Mexico and the American Southwest.

Much more extensively documented is the Norse discovery of America, recorded in Icelandic manuscripts of the Middle Ages. The Norse, or the Vikings as they came to be called, were ideally fitted to be wide-ranging explorers. Because of their geographical location, it was natural that they should be the first to discover territory in the Western Hemisphere. Their ships were built to cross the stormy seas of the North. The early Norsemen had become expert seamen while living on the Baltic Sea. There were Danish raids on England as early as the eighth century. During the next two centuries, they explored arctic areas and crossed the Atlantic. They colonized the Shetland Islands, the Orkneys, the Hebrides, the Faroes,

Iceland, and Greenland, conquered the Isle of Man, portions of northern Ireland and Scotland, central England and Normandy. The Norsemen were among the greatest seafaring people of their times. They had sailed past enemy European shores from the Arctic to the Mediterranean. Their ships were the best, and the sea held no terrors for them. Throughout the Middle Ages the Norsemen were spreading out east, west, and south. Having established themselves by conquest or colonization three-fourths of the way across the Atlantic, it was natural that they would find their way into the last quarter. As the historian John Fiske points out, "When once the Northmen had found their way to Cape Farewell, it would have been marvelous if such active sailors could long have avoided stumbling upon the continent of North America."

The original narrations of the early Norse voyages to America are contained in two collections of sagas, known as the *Flatey Book* and the *Hauk's Book*, both dating from the fourteenth century. Accounts in both state that the first discovery of America's mainland was accidental. The *Flatey Book* tells the story of this discovery by a Norseman named Bjarni Herjulsen, who describes the east coast of America, possibly including Cape Cod. This find was not followed up by the original discoverer, but by Leif Erikson, the son of Erik the Red. In 999, young Erik, then about age twenty, sailed out to see the world, making the first direct crossing of the Atlantic in history. The North American coastal region discovered by Bjarni was named Vinland (Wineland), because of the profusion of grapes. It was once identified by historians as New England, but is now believed to be northern Newfoundland.

The Norsemen first visited Iceland near the middle of the ninth century. The explorers were impressed by the rich fisheries and the pleasant green valleys along the shores. The settlers organized local units of government. The restless Norsemen began to look for other lands to explore and to possess. Erik decided to head an expedition to find the country that had been rumored to exist in the west. To attract settlers to the new land, Erik called it Greenland. He and his followers set up a colony in a district known as the Eastern Settlement. In another district on Greenland's western coast soon rose the Western Settlement. It was not an attractive area. Greenland has one of the world's largest glaciers, 1,500 miles long and almost 700 miles wide. The land is bleak and treeless, covered by perpetual ice except for a strip of land along the southwest side.

Despite the uninviting outlook, the Greenlanders carried on for 500 years. The two settlements were never populous. There were between 2,000 and 3,000 inhabitants living on about 280 farmsteads. The Greenland settlement probably existed until the fifteenth century. By that time its connections both with the mother country and its newly discovered lands had ceased. There is no evidence that Christopher Columbus ever knew about the early Norse discoveries or that they in any way influenced him.

The fate of the Greenland settlers is uncertain. Among the disasters they encountered were cattle diseases, epidemics, earthquakes, and famine. Relief supplies from Norway were few and far between because of the hazards of shipping. Emigration back to Norway reduced the population. One theory is that Greenland's Norsemen finally joined the Eskimos and went west, although that idea has not been authenticated.

Sagas are not the only clues to Viking explorations in America. In Nova Scotia and elsewhere in the northeastern part of the United States, Viking axes, boat keels, and other objects apparently of Viking origin have been found.

Wide popular interest in the Norse migration into North America apparently inspired a number of hoaxes. A celebrated specimen is the Kensington rune stone, emanating from southwestern Minnesota. While clearing land on his farm near Kensington, a Norwegian-American farmer, Olaf E. Ohman, turned up a stone weighing about 230 pounds. With the help of fellow-conspirators, the face of the stone was inscribed with runic characters. The "find" received wide attention, but was subsequently revealed as fraudulent. Another notorious example is the "Vinland" map, now in the Yale University Library. Supposedly a world map, Iceland, Greenland, and Vinland are shown as islands. Again, the map was unmasked as bogus, a clever hoax.

REFERENCES

Blegen, Theodore C. *The Kensington Rune Stone, on an Old Riddle*. St. Paul, Minn.: Minnesota Historical Society, 1968.

Christensen, T. P. *Discovery and Re-discovery of America*. Cedar Rapids, Iowa: Lawrence Press, 1934.

Gathorne-Hardy, C. M. *The Norse Discoveries of America*. Oxford: Clarendon Press, 1921.

Gordon, Cyrus H. *Before Columbus; Links between the Old World and Ancient America*. New York: Crown, 1971.

Holand, Hjamar R. *Explorations in America before Columbus*. New York: Twayne, 1956.

Redmond, Jeffrey R. *Viking Hoaxes in North America*. New York: Carlton Press, 1979.

CHAPTER

15

Amerinds — Origin

The history of American Indians began with a basic geographical misconception on the part of Christopher Columbus. Wishful thinking on Columbus's part had been greatly influenced by reading Marco Polo's *Travels*, which led him to believe that the world was smaller than it actually is, and that Asia extended east so far that it could be reached by a relatively short voyage westward across the Atlantic. Therefore, when Columbus discovered islands near the North American continent, he was convinced that he had reached the Indies and that the inhabitants were Indians. He was anxious to believe that he had found the Indies. The permanent impact of his ideas was the survival to the present day of such names as "West Indies" and "Indians."

Obviously, the natives encountered by Columbus were not East Indians. If not, who were they and how did they get to America? Answers to these questions have preoccupied the attention of historians, archeologists, and anthropologists for generations.

Over the centuries, a variety of explanations have sought to explain the origin of American Indians. Because of vague resemblances between American and ancient Egyptian monuments such as pyramids, it has been theorized that there was some contact between Egypt and America in former times, either directly by boat or indirectly by way of the mythical lost continent Atlantis. If so, Indians are of Egyptian origin.

Equally far-fetched was the claim stated by some sixteenth-century writers that Indians were the lost tribes of Israel. They attempted to show that the Indians resemble the Jews in character, dress, religion, physical characteristics, condition, and customs. In *The Book of Mormon*, Joseph Smith gives an historical account of the Israelites in America.

The Atlantis theory, which had considerable vogue, was based on the idea that the fabled island of Atlantis had existed somewhere in the Atlantic, constituting a land bridge to the New World, within possible reach of Europe. Subsequently a cataclysm caused Atlantis to sink beneath the ocean without leaving a trace.

Historical literature is full of accounts of discoveries in northeastern North America by such Vikings as Erik the Red, Leif Erikson, and others several centuries before Columbus. There is archeological support for some of these accounts.

All such attempts to explain the origin of the aboriginal American population have no validity in the light of modern knowledge. In any event, they left no cultural or physical effect on American Indians.

Anthropologists in general agree that Asia was the origin of the Indian and that most of the population came across the Bering Strait up to 40,000 years ago. The migration probably started around the end of the glacial epoch and continued for thousands of years. Convincing evidence of Asiatic origin is that the American Indians belong to Mongoloid stock. They usually have straight black hair, broad faces, shovel-shaped incisors, reddish or brown complexions, often prominent noses, and little facial or bodily hair.

The conclusion that the Bering Strait was the gateway by which the first inhabitants of the Western Hemisphere entered is based on several facts: first, the migrants were at a low state of technological development, incapable of making long sea voyages, and second, the distance between the American and Asiatic continents at the Bering Strait is short, about forty-five miles with islands at the midway point, making it the only practical route. Migration, it is believed, did not occur at any one time, but rather, there were waves of people crossing over a long period of time. Along with the human factor, various mammals invaded North America during Ice Age times and their remains have been found in Alaska and as far south as Mexico.

The timing of the migrations had to be coordinated with periods between ice ages. Geologists believe that an ice-free land route extended along the arctic coast to the MacKenzie River into what is now the United States as long ago as 20,000 years. It seems likely that an open land route, extending from central Asia into central North America, existed 20,000 years ago.

Fossilized bones of the American camel, extinct and living forms of musk ox, the giant sloth, mammoths, archaic bison, saber-toothed tigers, lions, and native horse have been found, nearly all of which subsequently disappeared.

The successive waves of migrant people were not all of the same stock and differed somewhat in physical characteristics. They probably were people from different points of origin, speaking different languages. They brought with them such cultural elements as bows, harpoons, spear throwers, clubs, spears, daggers of stone or bone, stone implements, bone awls, woven baskets and mats, domesticated dogs, and cooking with heated stones in vessels of bark, wood, or skin.

The Eskimo people were the last to move into the New World. Most estimates place the time of their migration at about 2,000 years ago. The Eskimos closely resemble the Mongoloid type, but there are certain physical differences probably resulting from mixing with earlier Indian strains who preceded the Eskimos into the arctic area.

Estimates of the aboriginal population are based on information supplied by explorers. For the entire hemisphere, figures range between 50 million and 75 million for the period around A.D. 1200, about equally divided between the two continents.

Wide differences in languages must have complicated communication among various tribes of Indians. An *Encyclopaedia Britannica* table lists fifty-eight language families north of Mexico, divided among six linguistic groups. Each was confined to a small geographic area and to a small population.

If the Aleuts and Eskimos are omitted, the principal Indian families of North America are:

1. The Athabascans, including many tribes of Alaska and north Canada, as well as Apaches, Navahos, and others in the United States.

2. The Algonquins, including Abnaki, Delawares, Crees, Chippewas, and various other historic tribes who flourished along the Atlantic slope from Labrador to Virginia and westward to the Rocky Mountains. This group possibly included Cheyennes, Arapahoes, and Blackfeet.

3. The Iroquois, inhabiting the Saint Lawrence Valley and New York State, including Cherokees, Mohawks, Senecas, and others.

4. The Sioux, comprising Omahas, Osages, Winnebagos, Crows, Catawbas, and others.

5. The Caddoan family, including Pawnees, Arickarees, Wichitas, Caddos, and others.

6. The Yuman family, found in the Colorado River Valley and California, including the Maricopa and other tribes.

7. The Shoshones, with whom are classed Utes, Comanches, and Hopi.

8. The Pueblo Indians of New Mexico and Arizona.

9. The Mexican tribes, with whom are classed Aztecs, Toltecs, and Nicaraguans.

10. The Maya of Yucatan and Central America.

Men wandered southward and probably reached southernmost South America more than 8,000 years ago. Classification of the South American tribes is more difficult than for the North American. The principal groups are those belonging to the mountainous district of the northwest, Indians of the Peruvian region, the Bolivian highlands, and the Pampean regions. Languages are equally diverse in character.

In some instances, Indian tribes reached an advanced state of civilization. The Incas in South America, starting about A.D. 500, had a highly developed agricultural economy and a complex system of land tenure, government, and law. Another remarkable state of culture was reached by the Mayas from southern Mexico and northern Central America, principally Guatemala. The Mayas' highly developed civilization reached its peak in the seventh and eighth centuries. Outstanding features included an advanced system of hieroglyphics (picture writing), a system of numbers that included zero, and an accurate calendar. Mayan astronomers calculated the movements of the moon and the sun. Other Mayan accomplishments included building gigantic stone temples, producing fine carvings and painted murals, weaving textiles of cotton, and making paper of tree bark. A third outstanding Indian people were the Aztecs, who dominated Mexico until the Spanish conquest in the sixteenth century. Their learning and skills were apparently borrowed from the Mayas and Toltecs. The Aztecs had an accurate calendar, a system of counting, and a kind of hieroglyphic writing.

They made pottery, wove cloth, carved stone sculptures, cut gems, and worked metals. Their master architects designed aqueducts, dikes, and huge stone pyramids topped by temples. Predecessors of the Aztecs were the Toltec Indians, who controlled most of central Mexico between A.D. 900 and 1200. The Toltecs had a calendar, an advanced form of picture writing, and paper made of tree bark. Their architects used stone pillars to support the roofs of large buildings and designed flat-topped stone pyramids as bases for temples.

Best known of the aboriginal groups in the American Southwest are the Pueblo Indians, who built the great cliff villages and the equally remarkable multiple apartment houses of adobe and stone masonry.

The noteworthy civilizations of the Mayan, Incan, and Aztec Indians were destroyed by the Spanish conquerors, as described in the next chapter, "Indian Civilizations before Columbus."

REFERENCES

Amsden, Charles A. *America's Earliest Man.* Los Angeles, Calif.: Southwest Museum, 1935.

Baity, Elizabeth R. *Americans before Columbus.* New York: Viking, 1951.

Driver, Harold E. *Indians of North America.* Chicago: University of Chicago Press, 1969.

Fowler, W. S. *Ten Thousand Years in America.* New York: Vantage, 1957.

Gladwin, Harold S. *Men out of Asia.* New York: Whittlesy, 1947.

Hopkins, David M., ed. *The Bering Land Bridge.* Stanford, Calif.: Stanford University Press, 1967.

Jenness, Diamond. *The American Aborigines.* Toronto: University of Toronto Press, 1933.

MacGowan, Kenneth. *Early Man in the New World.* New York: Macmillan, 1962.

Martin, Paul S. *Indians before Columbus.* Chicago: University of Chicago Press, 1975.

Stafford, Harry E. *The Early Inhabitants of the Americas.* New York: Vantage, 1959.

CHAPTER

16

Indian Civilizations before Columbus

Centuries before Christopher Columbus came to the New World, several notable Indian civilizations had developed, as previously noted. Of outstanding significance were the Aztecs, the Incas, and the Mayas. All were destroyed by invading Spaniards.

MAYAS

The Mayan Indians achieved a high state of civilization in Central America before the arrival of Europeans. Their area comprised the states of Chiapas, Tabasco, Campeche, and Yucatan, the territory of Quintana Roo in Mexico, the basin of Petén and adjacent highlands in Guatemala, western Honduras, and British Honduras. Three natural regions became the centers of Mayan civilization: the southern region, including the plateaus and mountains of eastern Chiapas and southern Guatemala; the central region, including Petén and the southern part of the Yucatán Peninsula; and the northern region, including the northern part of the Yucatán Peninsula. The total area is about 130,000 square miles, nearly three-fifths the size of France.

Mayan history dates from the beginning of the Christian era, though its primitive progenitors can be traced back 3,000 years. There are two principal periods: the Old Empire, from 320 to 987, and the New Empire, which extended from 987 to the Spanish conquest. Ecologically, the area occupied by the Maya is extremely diverse, ranging from snow-swept high volcanoes to parched deserts and rain-drenched jungles.

During the Old Empire, the Mayas reached a remarkable level of intellectual and artistic heights, hardly matched anywhere, even in the Old World. The Mayas were the only indigenous people in America to invent a true system of writing. Their mathematicians worked out a system involving the concept of zero, an intellectual achievement matched by only two other civilizations—India and Mesopotamia. Their astronomers calculated the length of the solar year with precision. The Mayan calendar was based on a sacred year of 260 days and a secular year of 365 days. Mayan knowledge of architecture enabled them to build great ceremonial centers of cut stone, including palaces, temples, astronomical observatories, and other structures, lavishly embellished with sculptured decorations. Mayan sculpture, painting, pottery, and the art of jade carving were even more advanced.

The Mayan people were dominated by religious beliefs as interpreted by the priesthood. The priests held a monopoly on writing, mathematics, astronomy, and religious art and architecture. Immense religious centers were built. The most important center appears to have been Tikal in Petén, the highest-known Mayan building. The second-largest center was Copán in western Honduras, home of the finest Mayan sculpture.

A cruel custom among the Mayas, especially prevalent from about 1000 to 1400, was human sacrifice, usually of young maidens, to placate the rain god Yum-Chac and persuade him to send rain. The center for this ceremony was the great pyramid in the temple at Chichén Itzá in the upper part of Mexico's Yucatán Peninsula.

Mayan civilization declined even before the coming of the Spanish. Possible causes included earthquakes, changes in climate, civil war, a breakdown in government, and economic problems. The Spaniards first reached the Yucatán Peninsula in 1517 and completed their conquest of the peninsula in 1546.

The conquering Spaniards proceeded to enslave the Mayas, with the justification of Christianizing them. Priests took control in the middle of the sixteenth century. The rulers were deposed, their religion banned, the Mayan priests executed, and their books burned. Eventually half a million Mayas were made slaves and every vestige of their culture was rooted out and destroyed.

Diego de Landa, archbishop of Yucatán, was responsible for the destruction of important Mayan records. In 1562 he collected as many Mayan books as he could find, labeled them "works of the devil," and ceremonially burned them in a public square in the town of Mani. Hundreds of volumes of Mayan history and legend were thus lost. Only three books survived the bonfire.

INCAS

The Incas were a small tribe in the Andean mountains near Titicaca until about 1100, when they occupied the town of Cuzco in central Peru and began to create an empire. By the early 1400s they had brought all neighboring tribes under their rule. At its height the Incan empire stretched for about 3,000 miles from present-day Colombia along the Pacific coast through Ecuador, Peru, and Bolivia into Chile. The capital and sacred city was Cuzco.

The world of the Incas comes near to being unique. There are six great upland basins between 8,000 and 13,000 feet in elevation, walled in by mountains, behind which loom the great snow peaks of the Cordilleras. In the south are three great volcanic masses rising to a height of 20,000 feet. Much of the area is subject to earthquakes.

It is not known for certain when man first began to inhabit these upland basins, but it may have been as early as 9,500 years ago. Traces of permanent habitation have been found as high as 17,500 feet. It was in the temperate zone of the high mountains, however, that the main Incan civilization emerged, only about a century before the arrival of the Spanish conquerors. The Incas never developed any form of writing, not even picture writing, and kept no record of the cultures that preceded them. Archeological evidence indicates that the Incas inherited building techniques, a system of government, and such arts as ceramics, design, and weaving from the Chimu and earlier cultures.

According to Incan tradition, the ruling dynasty began with the probably mythical Manco Capac in the twelfth or thirteenth century. The first genuinely historical figure was Inca Viracocka who reigned from 1347 to 1400.

The Incan empire has been described as the world's most successful benevolent despotism. Every peasant family, clan, and tribe had lands assigned to it. The assignments were revised from time to time to make sure each family could support itself. Along with their own land, the workers had to cultivate those of the church and state. Church lands supported the priesthood of the state religion, which was sun worship. The state lands supported not only the ruling Inca and his officials and soldiers, but also aged and crippled subjects and those who had lost their own crops through misfortune.

The Incan empire had no money, and taxes were in the form of labor. Each able-bodied man was required to put in a certain number of days per year at such labor as constructing roads and public buildings. Older men were assigned light work or exempted. Agricultural produce was stored in state granaries to be used by any part of the empire in time of drought or other hardship. In areas of little rainfall, an elaborate irrigation system was developed with networks of canals and ditches. Incan farmers were the first to cultivate white potatoes and other food plants. They terraced mountain slopes to conserve soil and used fertilizer. They also domesticated the llama and alpaca for their wool.

The Incan builders erected temples and strongholds using blocks of granite weighing as much as 100 tons each. Mortar was not used; instead, the blocks were fitted tightly together. Cities were connected by roads which efficiently linked parts of the vast territory. A string of mountain fortresses was built along the eastern frontier in the beginning of the fifteenth century.

The Incas spoke the Quechua language, as do their descendants today. In place of writing, they invented the *quipu*, a device consisting of cords of different lengths and colors which, by arrangement of knots, recorded accounts and events. The artistic skill of Incan craftsmen was shown in bone flutes and other musical instruments, gold and silver ornaments, copper and bronze utensils, hand-woven textiles, and pottery. The Incas also made considerable progress in astronomy and chronology.

The prosperous Incas possessed great quantities of gold, and it was their reputation for great wealth that attracted the Spanish under Francisco Pizarro. The invaders captured the capital in 1532 and 1533. After a farcical trial, the Spanish executed the Incan emperor Atahualpa. Pizarro's followers looted the empire and crushed a series of Incan uprisings. A generation later the Incas were a vanquished people. Those who escaped death or slavery became the ancestors of the Quechua Indians.

AZTECS

The Aztecs, known also as the Mexica, are one of the three famous pre-Columbian civilizations in America. What made it possible for these great civilizations—the Aztecs in the Valley of Mexico; the Mayas in Guatemala, Honduras, and Yucatán; and the Incas in Peru and Bolivia—to rise to the top? Alfonso Caso, in his work *The Aztecs*, concludes, "In each case the domestication of a staple food supply seems to have been a deciding factor; among the Aztecs and Mayas it was maize or corn and among the Incas the white potato."

The Aztecs dominated Mexico at the time of the Spanish conquest. They were, however, comparative latecomers. The Valley of Mexico had been occupied by Indian tribes for centuries, among whom the Toltecs were outstanding. According to tradition, the Aztecs migrated from the fabled land of Azatlan in the north toward the Valley of Mexico. By the time these northern invaders reached Mexico in the eleventh century, the powerful civilization of the Toltecs was in decline. The Aztecs arranged alliances with related tribes, spread their influence over a large area by force of arms, established their trade and religion, and made Tenochtitlán, now Mexico City, the center of their "empire." They held political sway over most of Mexico and as far south as Guatemala, El Salvador, and Nicaragua.

The Aztec conquests did not come easily. Some towns offered prolonged and bitter resistance. Those that paid voluntary tribute were regarded as allies. Rebellious towns had captives taken away for sacrifice.

When the Aztecs entered Mexico, the Toltecs already had a well-developed culture based on agriculture. The Aztecs did not originate, but borrowed and elaborated on what they found. Surplus food was provided by cultivation of the soil, and the development of arts and crafts began. Beautiful fabrics were woven from the cotton that they grew, and pottery of excellent quality was made. Major markets grew up and traders brought goods from a distance. The Aztec cities became great ceremonial and political centers, with gardens, large public buildings, palaces, and temples often perched high on pyramids. Prime features of the ceremonials were human sacrifices on a tremendous scale and ceremonial cannibalism.

The capital city, Tenochtitlán, was built on an island in Lake Texcoco, one of five small lakes in the Valley of Mexico. Its population has been estimated at several hundred thousand. The city was crisscrossed by hundreds of canoes bringing goods from other cities. Three raised roads led from the mainland to the capital's central plaza, where the largest public buildings, temples, and palaces were located. Nearby were a ceremonial ball court, a rack for the skulls of sacrificial victims, and one of the city's five markets. Aqueducts brought fresh water to the plaza from springs on the mainland.

Such learning and skills as the Aztecs possessed came mainly from the Toltecs, Mayas, and other earlier peoples of Mexico. The Aztecs had an accurate calendar, a system of counting by twenties, paper made from fig-tree bark, and a system of picture writing. Their craftsmen made pottery, wove cloth, carved stone sculpture, cut gems, and worked metals. Master architects designed aqueducts, dikes, and huge stone pyramids topped by temples.

The Aztecs worshipped many gods, the most important of whom were the Hummingbird Wizard, war god, sun god, and rain god. The Aztecs believed that their gods needed human hearts and blood for nourishment. To get captives for sacrifices, they waged war almost incessantly. Priests sacrificed victims by cutting open their chests and pulling out their hearts. On one occasion 20,000 victims were sacrificed. Caso asserts in *The Aztecs* that "their entire existence revolved around their religion and there was not a single act, public or private, that was not tinged by religious sentiment."

In 1519 the Aztec empire was invaded by a small Spanish army led by Hernando Cortez. The Spanish conquered the Aztecs because a great majority of people simply watched the conflict or actively aided Cortez. Cortez gained allies among tribes resenting Aztec power. Montezuma II, the Aztec emperor, welcomed Cortez to Tenochtitlán, but was seized as a hostage by the Spaniards. The emperor was killed by his own people who showered him with stones and arrows. The Spaniards gradually reduced the Indians to serfdom. Many lost their land and were forced to work in mines. Others died of tuberculosis, measles, and smallpox. By 1521, the Aztec empire had come to an end.

REFERENCES

Mayas

Bennett, Evan. *The Maya Epic.* River Falls, Wis.: University of Wisconsin—River Falls Press, 1974.

Coe, Michael D. *The Maya.* London: Thames & Hudson, 1966.

Craven, Roy C. *Ceremonial Centers of the Maya.* Gainesville, Fla.: University of Florida Presses, 1974.

Livesey, Herbert B. "Mayan Mexico." *Travel & Leisure* 17 (February 1987): 78-91, 124-32.

Pearce, Kenneth. *The View from the Top of the Temple.* Albuquerque, N. Mex.: University of New Mexico Press, 1984.

Rivet, Paul. *Maya Cities.* New York: Putnam, 1960.

Incas

Blasingame, Wyatt. *The Incas and the Spanish Conquest.* New York: Messner, 1980.

Hanstein, Otfrid von. *The World of the Incas.* Freeport, N.Y.: Books for Libraries Press, 1925.

Hemming, John. *Conquest of the Incas*. New York: Harcourt, 1970.

Innes, Hammond. *The Conquistadors*. New York: Knopf, 1969.

Prescott, William H. *History of the Conquest of Peru*. London: Bentley, 1847.

Aztecs

Caso, Alfonso. *The Aztecs: People of the Sun*. Norman, Okla.: University of Oklahoma Press, 1958.

Davies, Nigel. *The Aztecs, a History*. New York: Putnam, 1974.

Prescott, William H. *History of the Conquest of Mexico*. London: Bentley, 1843.

Vaillant, G. C. *The Aztecs of Mexico*. Garden City, N.Y.: Doubleday, 1944.

CHAPTER

17

Machu Picchu

One of the great archeological discoveries of the twentieth century was made in 1911 by one-time U.S. Senator Hiram Bingham, high in the Andes of Peru. This, the last stronghold of the Incas after the Spanish conquerors overthrew the Incan empire in the sixteenth century, had been abandoned and lost for centuries. Known today as Machu Picchu, its Incan name is unknown. It is located fifty miles northwest of the ancient Incan capital of Cusco, 2,000 feet above the Vilcanota River on a narrow ridge between 10,300-foot Machu Picchu ("old peak") and 9,000-foot Huayna Picchu ("new peak").

A dramatic account of the search for and rediscovery of Machu Picchu is told by Bingham in his *Lost City of the Incas; The Story of Machu Picchu and Its Builders*. The question facing Bingham and the Yale Peruvian Expedition, which he headed, was whether the young Incan emperor Maneo, fleeing from Pizarro's armies, had established a new capital in the wilds of Vilcabamba. It was Bingham's mission to verify vague rumors he had heard of ruined palaces and temples in the wilderness.

The approach to Machu Picchu is graphically described in Bingham's account. Following the Urubamba Valley, he found that the Urubamba River had cut its way through precipitous cliffs. This was the natural gateway to the ancient province of Vilcabamba in which Machu Picchu is located. For centuries it had been virtually closed by the combined efforts of nature and man. Dangerous rapids on the river were impassable, but the precipices could be scaled with considerable effort. Bingham was vastly impressed by the variety and extent of the valley's scenery. "Not only has it snow-capped peaks, gigantic precipices of solid granite rising abruptly thousands of feet from its roaring stream, and the usual great beauty of a deep canyon winding through mountains of almost incredible height, but there is added to this the mystery of the dense tropical jungle and the romance of the ever-present remains of a bygone race." The river was crossed a few inches above the roaring rapids by a primitive bridge made of four logs bound together with vines.

Climbing farther up the ridge, Bingham and his native helpers suddenly found themselves in the midst of a jungle-covered maze of small and large walls, the ruins of buildings made of blocks of white granite. They realized that they had discovered Machu Picchu.

A group of Indians who had been living among the ruins for several years were growing sweet and white potatoes, corn, sugar cane, beans, peppers, tomatoes, and a kind of gooseberry. These surviving Incas had found a practically impregnable place of refuge.

Located on a ridge that ends in a magnificent peak, the Machu Picchu ruins are flanked by precipices on both sides and many terraces evidently intended for agricultural purposes. There are also stone-lined watercourses and three small springs. It is likely that an irrigating ditch was dug many miles back into the mountains to assure a larger supply of water.

Machu Picchu was lost for centuries because the ridge on which it was built is in the most inaccessible corner of the most inaccessible section of the central Andes. No section of the Andean mountains is better defended by natural bulwarks—a stupendous canyon of granite stone, with a sheer, three-thousand-foot precipice.

One of the most conspicuous features of Machu Picchu is the stairways, more than a hundred in number, large and small, ranging from three or four steps to as many as 150. The longest stairway served as the central thoroughfare in the heart of the city. A peculiar aspect of the buildings is that each has only three sides, with the fourth side entirely open and looking out over the canyon to the rising sun.

Bingham concluded that the "Lost City of the Incas," as he termed it, had a history that spanned many centuries. Once called Vilcapampa, it is now named for the region where it is located. Unlike most of the Incan empire, Machu Picchu was never looted, and it has been a vital source of information about the Incas.

In its lost state the city was a carefully guarded treasure house where the worship of the sun, the moon, the thunder, and the stars was restored. The last four Incan emperors had their safest and most comfortable home there, and the Chosen Women, or Virgins of the Sun, whose lives had been devoted from early girlhood to all the duties of the sanctuary, found a refuge from the animosity and lust of the conquistadors. On ceremonial occasions and during great festivals, a gay procession of Inca nobles, priests, and Virgins of the Sun probably marched up the great stairway to the top of the hill to witness the worship of their favorite deity—the sun god. Because the bones found among the ruins seemed to be chiefly those of young women and old men, scholars have concluded that the city was primarily a hideaway for Virgins of the Sun and priests.

At the center of Machu Picchu are well-built dwellings, believed to have been for the Priests of the Sun. A circular temple over the great ledge was probably the center of sun worship. Beyond the stairway is a magnificent compound, likely the residence of the Incan emperor himself.

On the north side of Machu Picchu are the remains of hundreds of ancient agricultural terraces built to provide the inhabitants with such foods as potatoes, corn, beans, tomatoes, and sweet potatoes. A small conduit brought the water supply along the terraces and behind the temple to the Stairway of the Fountains. The Gateway of the Sun was the largest of several gates. It was chiseled from a single block of hard andesite, was 10 feet high and 12.5 feet wide, and weighed nearly ten tons.

A signal station was located on top of Machu Picchu to detect the approach of enemies or unwelcome visitors. The station is at the edge of one of the most stupendous precipices in the Andes—three thousand feet to the bottom.

The question remains whether Machu Picchu was meant to be a fortress, a kind of walled city. There is no single continuous wall encircling the entire structure. However, because of its location, almost impregnable against any potential attackers, the city hardly needed defensive walls of any kind.

Bingham led additional expeditions to Machu Picchu in 1914 and 1915 to clear the site and further explore the ruins. Machu Picchu is presently being restored. Because of its remarkable architecture and spectacular setting, it has become a prime attraction for archeologists and tourists.

REFERENCES

Bingham, Hiram. "The Discovery of Machu Picchu." *Harper's Magazine* 126 (April 1913): 709-19.

Bingham, Hiram. *Lost City of the Incas; The Story of Machu Picchu and Its Builders*. New York: Duell, Sloan and Pearce, 1948.

Bingham, Hiram. *Machu Picchu, a Citadel of the Incas.* New Haven, Conn.: Yale University Press, 1930.

De Camp, L. Sprague. "The Lost City of the Incas." *Science Digest* (August 1967): 68-72.

Von Hagen, Victor W. *A Guide to Machu Picchu.* New York: Farnam, 1949.

CHAPTER

18

Raleigh's Lost Colony

Preceding the first two successful attempts to establish English colonies in America—the Jamestown settlement under Captain John Smith and his associates in 1607, and the arrival of the *Mayflower* at Plymouth in 1620—a group of 112 English men, women, and children settled in 1587 on Roanoke Island in North Carolina. Their goal was to establish the first permanent English colony in the New World.

The moving spirit behind the Roanoke venture was Sir Walter Raleigh, a favorite of Queen Elizabeth, who had authorized Raleigh to explore and take possession of the area north of Spanish Florida.

Raleigh's first expedition, sent out in 1584, consisted of two vessels under the command of Captain Philip Amadas and Arthur Barlowe. They left England in April and reached the Outer Banks of the North Carolina coast in July, named the new land Virginia in tribute to the Virgin Queen, and chose Roanoke Island as a possible site for a future settlement. The small band of men explored the region as best they could, traded with the Indians, and observed plants, trees, animals, and Indian customs.

Amadas and Barlowe returned home with glowing reports on the veritable paradise they had discovered, the friendly Indians, the plentiful fruits and fish. Two natives, Manteo and Wanchese, accompanied the expedition back to England.

In fact, the location chosen by Amadas and Barlowe had serious natural disadvantages. The coast of North Carolina is a long, narrow chain of low sand hills that separate the ocean from two broad, shallow bodies of water—Pamlico and Albemarle sounds—which are the estuaries of the Neuse, the Roanoke, and other great rivers in the state. At irregular intervals the line of the Banks, as the coast is locally known, is broken by narrow and ever-shifting inlets. Through them flow the ocean tides, turning the inner waters into vast salt lakes. Within the barren downs are a few small islands, among them Roanoke. Roanoke Island is about twelve miles long and three miles wide, some six miles from the mainland. There is no natural harbor. The area is noted for stormy weather and shipwrecks.

Inspired by the enthusiastic reports brought back by his agents, Raleigh sent out a second expedition in the spring of 1585. Under the command of Sir Richard Grenville, the flotilla consisted of seven vessels and some 600 men, about half of whom were soldiers. The two Indians, Manteo and Wanchese, accompanied them. After surveying the surrounding area, Grenville went back to England in August, leaving 107 men on Roanoke Island to build houses and a fort.

The detachment that remained on Roanoke Island through the winter of 1585-1586 was under the command of Ralph Lane, a soldier. Lane spent most of his time searching for gold in the sound and rivers to the west, depending on the soldiers to provide food. He did not trust the Indians and considered them to be savages and barbarians. After Lane ordered a surprise attack on an Indian village that killed a chief and several other Indians, the Roanokes refused to give further help to the Englishmen. By spring Lane's forces were suffering from an acute shortage of food.

In June, Sir Francis Drake visited Roanoke Island on his way back to England following successful raids against the Spanish in Florida and the Caribbean. When he heard of Lane's problems, Drake offered to leave supplies. While anchored off the coast, however, Drake's fleet was scattered by a storm. Lane was ready to give up and decided to abandon the settlement and return to England with Drake.

Lane had quit too soon, for a relief vessel with supplies sent out by Raleigh was near the coast. When it found the settlement abandoned, the ship returned to England. Three weeks later, a fleet of three vessels commanded by Grenville also arrived at Roanoke Island. Grenville left fifteen men to hold the fort during the winter of 1586-1587.

Back in England, Raleigh was not ready to abandon his dream of establishing a permanent settlement in America. He decided to launch a third expedition. The fleet of three vessels left England in early May 1587, with 112 passengers on board, among them seventeen women and nine children. Raleigh ordered John White, the governor of the colony, merely to stop briefly at Roanoke to pick up the men left by Grenville and then to proceed to Chesapeake Bay. At Roanoke, Captain Simon Ferdinando, commander of the fleet, refused to proceed any further and compelled White, much against his will, to land his colony on the island. The only trace they found of Grenville's men were their bones. Apparently, all had been slain by the savages.

On August 18, 1587, Governor White's daughter, Eleanor Dare, gave birth to a daughter. Virginia Dare was the first white child born in America. On August 27, Governor White returned to England for supplies. From that date to the present, the fate of the Roanoke colonists has been a mystery.

The plan had been for White to return in the spring of 1588, but by then England was at war with Spain. Every English ship and English seaman was in demand to confront the "Invincible Armada." Sir Walter Raleigh, however, fitted out a small expedition for the relief of his colony. Two vessels sailed for Roanoke in April 1588, but were soon driven back by Spanish war vessels.

It was nearly two years later, in March 1590, that White finally sailed. Even then he did not command a ship, but was a passenger on a vessel engaged in the West Indian trade. He reached Roanoke in August and found the island abandoned and everything in ruins. Before leaving Roanoke, White had instructed the colonists that if they changed their residence during his absence to carve on a tree the name of the place where he could find them. On one of the door posts of the fort he found the word *CROATOAN*. White assumed that the colonists had moved to Croatoan, where Manteo had been born. Preparations were made to proceed to Croatoan, but a storm came up, and food and fresh water were low. The captain, who was anxious to proceed to the West Indies to search for Spanish treasure ships, refused to continue the search for the colony. Later, no fewer than five expeditions were sent out to find the colonists, but without success. By the time the last one returned in 1603, Raleigh was in prison and his private funds, amounting to £40,000 were exhausted.

Various theories have been offered to account for the disappearances of the Roanoke Island colonists 400 years ago. One belief is that all were massacred by hostile Indians. A similar theory is that they were killed by the Spanish, but a search of Spanish archives revealed no such happening. One historian suggested that the colonists finally gave up hope for relief from England, built a small vessel, and were lost at sea while attempting to sail across the Atlantic.

Local legends in North Carolina maintain that the lost colonists survived and intermarried with the Indians, and that their descendants live in the region today. There is a tradition among the Pembroke Indians still living in Robeson County in southeast North Carolina that they descended from the lost colony. A similar tradition has been reported by the Haliwar Indians, who live near the present-day Carolina-Virginia border in Halifax and Warren counties. All the evidence points to the theory that the colonists abandoned their settlement on Roanoke Island and moved to Croatoan, now known as Hatteras Island. That is the conclusion of Pulitzer-Prize-winning dramatist Paul Green in his symphonic play *The Lost Colony,* performed each summer for some years on Roanoke Island.

Some attempts were made by later colonists at Jamestown, Virginia, to find the Roanoke colonists and to verify rumors received from the Indians, but nothing conclusive was ever learned.

Supporting evidence for the theory that the Roanoke colonists joined the aborigines is the discovery of a tribe of gray-eyed Indians in Robeson County, about 200 miles from Roanoke. Known as the Lumbee Indians, they have complexions ranging from dark to very fair. Blond hair and blue eyes are not uncommon. Of the ninety-five surnames of the "lost colonists" of Roanoke—names such as Sampson, Cooper, and Dare—forty-one can be found among the Lumbees.

REFERENCES

Durant, David N. *Raleigh's Lost Colony.* New York: Atheneum, 1981.

Kupperman, Karen O. *Roanoke, the Abandoned Colony.* Totowa, N.J.: Rowman, 1984.

Powell, Williams. *Paradise Preserved.* Chapel Hill, N.C.: University of North Carolina Press, 1965.

Stick, David. *Roanoke Island, the Beginning of British America.* Chapel Hill, N.C.: University of North Carolina Press, 1983.

Weeks, Stephan R. "The Lost Colony of Roanoke." *Papers of the American Historical Association* 5 (October 1981): 413-46.

CHAPTER

19

North Pole Discovery

Few men in history have shown such single-minded dedication to achieving one objective as Robert Edwin Peary. Early in his career he determined to become an arctic explorer, with the ultimate aim of reaching the North Pole. As he expressed himself to Theodore Roosevelt in 1906 after the failure of his next-to-last expedition, "To me the final and complete solution of the Polar mystery—is the thing which should be done for the honor and credit of this country, *the thing which it is intended that I should do, and the thing that I must do!*"

Numerous arctic explorers preceded Peary in attempts to reach the North Pole. No region of the earth has produced more tales of heroism and stoic endurance, of narrow escapes and tragedies, than the Arctic. Examples are John Davis, one of the most remarkable of Elizabethan seamen, who in 1588 reached a point 1,128 miles from the Pole; and, several centuries later, David Buchan, another Englishman, who in 1818 sailed with instructions to reach the Sandwich Islands by way of the North Pole and to return, if possible, by the same route. Far short of his goal, Buchan concluded that the task he had been assigned was hopeless.

Other nineteenth-century expeditions—by Ross, Austin, Franklin, McClure, Greely, Nares, Markham, and Nansen—followed each other in rapid succession. It was finally conceded that there was no northern passage across the Pole, after which the prime aim became attainment of 90 degrees north latitude, that is, the Pole itself, a goal that required nearly a century to achieve. All the explorers prior to Peary had to admit defeat and many failed to return. A noteworthy effort was made by Sir William Edward Parry in 1827 and described in his book *Narrative of an Attempt to Reach the North Pole in Boats Fitted for the Purpose*, published in London in 1828. Parry tried to reach the pole from Spitsbergen using boats shod along their keels with iron strips and dragged by British sailors; on ice the boats were sledges, on water they became boats again. The expedition was still 445 miles short of the Pole when Parry abandoned the struggle. He attained 82° 45' north latitude.

A new record was established in 1875 by Sir George Nares, who went three miles farther north than Parry. Nares was followed by J. B. Lockwood and D. L. Brainard of A. W. Greely's U.S. Army expedition, who exceeded Nares's record by four miles in 1882. In the winter of 1895 Friftjof Nansen and Frederick Hjalmar Johansen sledged northward to 86° 12', breaking the Greely record by 186 geographical miles; five years later, a party of Italians under the Duke of Abruzzi sledged from a land base in Greenland Sea, East Greenland, to 86° 34'.

Robert Peary poured a lifetime of rugged conditioning and experience into preparation for his drive to the North Pole. He had grown up in Maine, received an engineering degree from Bowdoin College in 1877, and became a naval officer in 1881. During the years 1884-1888 he was a part-time engineer-in-chief of a Nicaraguan canal survey. He went on his first arctic expedition to Greenland while still on leave from the Navy in 1886. From 1891 until the end of his naval career he was engaged in arctic explorations more or less continuously.

Peary's second northern expedition, 1891-1892, established headquarters at McCormick Bay on the western coast of Greenland. From there, he made extensive sledge excursions, ascended the summit of the 5,000-8,000-foot high great ice cap which covers the interior of Greenland, pushed northward for 500 miles over a region previously unvisited by any white man, and on July 4, 1892, discovered Independence Bay. A broken leg was Peary's only injury on these strenuous travels.

Peary soon became convinced that Eskimos had much to teach the white man in the matter of polar expeditions. In this important respect, his work was revolutionary. Without the Eskimos, it is extremely doubtful that Peary could ever have reached the North Pole. In early polar expeditions, travel was undertaken during the summer because of dread of the northern cold winter. William Edward Parry's men, for example, had practically hibernated, digging themselves in during the fall and emerging in the spring. It is known that the Eskimos drove their dogs and sledges over the snow, but Parry waited for the snow to melt and then dragged carts through the mud.

Robert Peary rejected these ideas. He completely adopted Eskimo ways. Instead of traveling in summer he made full use of winter, insisting that serious northern journeys should begin in midwinter and end before the spring thaws. Instead of carrying heavy and burdensome tents, Peary built snow houses or igloos, Eskimo fashion. When trained in the art, four men could build an igloo in an hour's time, using strong snow-sawing knives. Another advantage of the snow houses: on the return journey an explorer could usually find the igloos built on the outward journey still intact.

Peary's sledges were also of the Eskimo type, though built with superior materials and lengthened to make them safer for crossing breaks in the ice or for traveling on thin ice. Hauling power was supplied by Eskimo dogs working in Eskimo harnesses, hitched in the Eskimo manner. Furthermore, Peary's clothes were of Eskimo cut, made of local skins, and worn Eskimo style.

Food was a crucial consideration for Peary, as it had been for all arctic explorers. Supplies of fresh meat were available from caribou, arctic deer, musk-oxen, and occasional polar bears. Certain basic provisions necessary for survival, however, had to be carried along, and on this subject Peary held strong views: "On a serious northern sledge journey," he declared, "only four tried articles—pemmican, tea, condensed milk, and hardtack—are necessary. Of all the items which go to make up the list of supplies for a polar expedition, the one which ranks first in importance is pemmican. With pemmican, the most serious sledge journey can be undertaken and carried to a successful issue in the absence of all other foods."

Pemmican, as made up for Peary's expeditions, was an American Indian invention consisting of lean meat dried, pounded fine, and mixed with melted fat, a mixture to which Peary added a few raisins per pound for flavoring. Milk and sugar in hot tea were also used for flavor and variety, rather than dietetic value. A pound of biscuit or hardtack per day was considered important for bulk.

At age fifty-two, Peary realized that his seventh attempt to reach the North Pole would probably be his last. His preparations for the expedition, therefore, were made with extraordinary care. For a 1906 try, he had designed and constructed the *Roosevelt*, a boat built to withstand the crush of masses of ice. The *Roosevelt* was a three-masted steamship, built entirely of white oak, with walls up to thirty inches thick. Her heavy bow was backed by twelve feet of solid deadwood, and the stern was reinforced by iron and had a long overhang to protect the rudder from ice. The rudder itself could be lifted out of the water when jammed or entangled. The ship was completely overhauled in readiness for the 1908 expedition.

In July 1908, the Peary party sailed from New York. At Etah, an Eskimo settlement in northwest Greenland, a number of Eskimos were taken aboard. The tribe from which these natives came was known variously as the Whale Sound tribe, the Polar Eskimos, or the Arctic Highlanders. Their dialect, customs, and traditions were somewhat different from Eskimos in other areas, an indication of an isolation of many centuries. There were about thirty-five families in the tribe and all were well known to Pearty through previous contacts. He was also familiar with their language.

The *Roosevelt* pushed steadily northward, through the Kennedy and Robeson channels, the vessel picking its way between ice floes. The voyage was hazardous in the extreme. There was a strong tidal current, which created a moving pack of chaotic ice floes and icebergs in midstream. For days the channel was frozen solid, and the *Roosevelt* had to find a sheltered cove until the way was again clear. At times gigantic floes, an acre or more in size, surged toward the ship, threatening to crush it. A charging iceberg smashed the bulwark. But three weeks later the Arctic Ocean came in sight. Peary's aim was to reach Cape Columbia on the north coast of Grant Land; the ice pack, however, prevented any progress beyond Cape Sheridan. On the first day of September the *Roosevelt* anchored in a snug harbor under the protection of Cape Sheridan, and the crew went into winter quarters. Not until February, when the first gray light of approaching dawn began to break through the darkness, could the dash to the Pole begin. Several hunting trips were arranged to pass the time and to provide the party with a plentiful supply of fresh meat during the winter.

On February 28 and March 1, 1909, Peary's men divided into several groups and set out. The party was composed of twenty-four men (six whites, one Negro, and seventeen Eskimos) and 133 dogs. All headed straight north for the Pole, 413 nautical miles away. Advance groups broke trail, while Peary brought up the rear. The greatest problem throughout the arduous trek, and the cause of frequent delays, was "leads," channels of water running through the ice, which had to be crossed with the sledges. For lack of supplies, the last trailbreaking group was sent back to the base camp on April 1. That left Peary with a supporting group of five men for the final drive—four Eskimos and Matthew Henson, a Negro, his long-time valet and companion, who was also a skilled dog driver. The best dogs, forty in number, had been reserved for the final march. They pulled five sledges. All were in excellent condition.

As the small party pushed north it continued to encounter many cracks in the ice. The sledges had to charge across from one ice cake to another as though leaping across a series of swaying stepping-stones. On one occasion Peary fell in the water but, since his clothes were watertight, he scraped off the ice and proceeded. Despite the obstacle course they were pursuing—open leads and massive hummocks of ice to avoid—the party was able to travel fast. The distance covered during the last five marches averaged twenty-six miles daily.

The Pole was reached on the morning of April 6, 1909. To make certain of the exact location and to allow for any slight error in computation, different observations of the sun were taken and marches were made ten miles in every direction. Afterward, flags were raised on a heap of snow and a record was left of the expedition. Peary spent thirty hours at the pole and photographed the desolate, snow-swept scene in all directions. The return journey was begun after a short sleep in an igloo.

Of scientific interest are the soundings that Peary made in the polar area. Five miles from the Pole itself, the depth was found to be 1,500 fathoms, or 9,000 feet, without touching bottom. A sounding in 1937 reached bottom near the Pole at 13,200 feet. The Pole, as discovered by Peary, lies in an ocean, 391 geographical miles from the nearest land.

The extraordinary speed that Peary made in reaching the Pole was exceeded in the journey home. "Our return from the Pole was accomplished in sixteen marches," Peary wrote. "It had been, as a result of our experience and perfected clothing, an amazingly comfortable return, as compared with previous ones." In order to avoid the danger of the spring tides, which would cause many breaks in the ice, Peary had decided the return must be speeded up and he planned double marches for the entire return trip. A continuous threat was the possibility that a gale might break up the ice, separating the floes by stretches of open water, and accelerate the drift eastward toward the Atlantic, melting the solid ice from beneath the explorers. On April 23, 1909, Peary and his party were back at Cape Columbia. One of the Eskimos was heard to mutter in his own language, "the Devil is asleep or having trouble with his wife, or we should never have come back so easily."

A few days later Peary reached the *Roosevelt* at Cape Sheridan. There he received the sad news that a young engineer, Ross Marvin, had broken through thin ice on the return to Cape Columbia and drowned—the only casualty during the entire expedition.

The *Roosevelt* reached the Labrador coast in early September. A message was relayed from there to Peary's sponsors, the Peary Arctic Club: "Stars and Stripes nailed to the pole. Peary." When the message was sent, Peary was unaware that a week earlier the triumph that he was announcing had been claimed by another American. Dr. Frederick A. Cook, who had once been associated with Peary, asserted that he, travelling with two Eskimos, had reached the Pole on April 21, 1908, a year ahead of Peary. Cook spread his story over the civilized world before Peary's message was sent. Many believed him and he was given receptions, balls, the freedom of cities, and other honors.

Peary reacted in fury. In several telegrams to news media, he denounced Cook as a liar, basing his statement on what Cook's own Eskimos had said, that they had never been out of sight of land. In a dispatch to the *New York Times*, Peary stated: "Cook has not been to the Pole on April 21, 1908, or at any other time. He has simply handed the public a gold brick." Almost all authorities soon came to agree that Cook's story was a hoax, because he was unable to offer any scientific evidence to support his claim. But Cook was a popular favorite, and Peary was viewed as arrogant and unsportsmanlike. Some persons even challenged Peary's own account of having reached the Pole. The controversy dragged on until the beginning of World War I, when more serious matters diverted the public's attention. The whole affair, unfortunately, had the effect of tarnishing Peary's great achievement and cast a shadow over the remainder of his career.

Racial prejudice was involved in one criticism of Peary. Why had he taken a black and four Eskimos on the last lap of his race and no white men? The answer was obvious in the case of the Eskimos. In defense of Matthew Henson, Peary pointed out: "In this selection I acted exactly as I have done on all my expeditions for the past fifteen years. He has in those years always been with me at my point farthest north. Moreover, Henson was the best man I had with me for this kind of work, with the exception of the Eskimos, who, with their racial inheritance of ice technique and their ability to handle sledges and dogs, were more necessary to me as members of my own individual party than any white man could have been." Also important was Henson's influence with the Eskimos; he persuaded them to travel over the heaving and splitting ice for the first time in the history of the tribe and prevented a mutiny on the way to the Pole.

Another Cook exaggeration, his claim to have ascended to the peak of Mt. McKinley in Alaska in 1906, was proved false. Fourteen years after the North Pole imbroglio, Cook suffered a greater defeat. In 1923, at Fort Worth, Texas, he was sentenced to more than fourteen years in prison for mail fraud in connection with an oil promotion.

A comparison of Peary's expedition to the North Pole and Roald Amundsen's conquest of the South Pole illustrates several important differences in exploratory techniques. In Peary's view, the South Pole was relatively easier to reach. Among significant dissimilarities pointed out by Peary are that the Arctic is a sea encompassed by lands, while the Antarctic is a continent surrounded by an ocean; the antarctic pole lies on a plateau two miles high while the arctic pole is in a sea two miles deep. A further handicap for the arctic explorer, stressed Peary, is that in trying to reach the

North Pole the traveler must sledge at the coldest time of the year in order to take advantage of frozen-over water, while in aiming for the South Pole travel is at the warmest period of the year when the ice is firm, resting on land, and there is no open water. Another great advantage of southern expeditions, Peary stated, is the food supply. Going south, caches can be set up on the forward march and picked up when homeward bound. In the Arctic, on the other hand, the northward journey is over moving ice and the possibility of finding a deposit on the return march is remote. There is also the likelihood of having a depot plundered by polar bears and foxes, for "a polar bear can and will tear asunder practically any sort of package." An antarctic depot is completely safe, however, for three reasons: it cannot float away because the land is stable; it cannot spoil, for there is neither rain nor decay; and it cannot be destroyed by wildlife, for there is none.

Amundsen, the discoverer of the South Pole, would doubtless have maintained that the problems of antarctic exploration might be different in kind but not in technical difficulties from those encountered in the arctic region. In exploration of the South Pole, there was frequent danger of losing a sledge with its load and men into a bottomless crevasse, although Peary held that this hazard was considerably less than the risk of losing life or load in the turbulent movement of the drifting arctic ice pack.

The great arctic explorer Vilhjalmur Stefansson commented on this issue: "Peary's forecast of the relatively easy attainment of the South Pole was destined to be confirmed within three years. It took hundreds of years to find the Antarctic mainland, but it was only sixteen years (1895-1911) from the first placing of human feet upon the shore of the southern continent until the first men stood at the southern Pole."

Perhaps no better tribute to Peary and his achievements can be found than was written by Stefansson: "Those before Peary pursued explorations with a heroic frame of mind. They did not set out to adapt themselves to difficulties; their whole purpose was to do or die in their attempt. Peary first set out to find how he could overcome the conditions of living in the Arctic. His great contribution to polar discovery lay in the fact that he introduced common-sense methods."

In belated recognition of Peary's feat, two years after the North Pole was attained, the U.S. Congress in 1911 adopted the following resolution:

> Your committee believe that in view of his long distinguished service in the Arctic regions in ascertaining the northern boundaries of Greenland, his soundings and tidal observations; his ascertainment of facts concerning the Arctic Ocean; the general information he has obtained by living over twelve years within the Arctic Circle; and finally having successfully followed a carefully

> laid out plan resulting in his reaching on April 6, 1909, and bringing back to civilization the conditions existing at the North Pole, that Robert Edwin Peary has performed a most remarkable and wonderful service, that he has attracted the favorable attention of the civilized world; and that, therefore, the American people, through its Congress, shall render him thanks, and bestow upon him the highest rank of the service which he adorns.

Peary was retired in the same year as a rear admiral. He became an aviation enthusiast and predicted the coming importance of air power in warfare. After his death in 1920, Peary Land, a peninsula in northern Greenland, was named for him.

The foregoing account of the discovery of the North Pole is widely accepted as true. The standard encyclopedias and histories of polar exploration almost invariably give Robert Peary credit for being the first explorer to reach the North Pole. Nevertheless, there are skeptics who have raised persuasive doubts. One of these is Theon Wright, author of *The Big Nail: The Story of the Cook-Peary Feud.* Wright made an exhaustive evaluation of the primary documents in the case, including all available records left by Peary and the extensive congressional hearings held prior to Peary's promotion to the rank of rear admiral. Wright questioned Peary's claim to have discovered the North Pole on three principal grounds: first, the failure to take longitudinal bearings; Peary conceded that he did not make these observations, regarding them as "a waste of time." Actually, it appears that Peary was not an experienced navigator. A second point was the rates of speed Peary claimed to have made over the ice—perhaps three times what would have been normal. Third, no account was taken by Peary of the eastward drift of the ice, which would have taken him off course and affected speed.

Wright concludes, "Evidence can be piled on evidence demonstrating the impossibility of Peary having reached the Pole on his 1909 journey. When his rate of speed is taken in conjunction with his lack of accurate observations and the factor of drift of the ice, the story becomes pure fiction." As a clincher Wright adds, "Whatever may be the verdict of historians with respect to Cook, the conclusion as to Peary's claim is inescapable. The reasons for the vague and indefinite research that has pursued this question over the years may be difficult to understand, but one thing is beyond question: the perpetuation of the myth that Peary discovered the North Pole has no possible justification in fact or tradition."

Some years elapsed before an uncontested discovery of the North Pole was attained. This was done by the crew of the U.S.S. *Skate*, an atomic submarine, when it surfaced at the Pole on August 8, 1958, thirty-eight years after Robert Peary's death.

REFERENCES

Angell, Pauline. *To the Top of the World.* New York: Bantam Books, 1966.

MacMillan, Donald P. *How Peary Reached the Pole.* New York: Houghton, 1934.

Mirsky, Jeanette. *To the Arctic.* Chicago: University of Chicago Press, 1970.

Rawlins, Dennis. *Peary at the North Pole, Fact or Fiction?* Washington, D.C.: Luce, 1973.

Stefansson, Vilhjalmur. *Unsolved Mysteries of the Arctic.* New York: Macmillan, 1939.

Weems, John E. *Race for the Pole.* New York: Holt, 1960.

Wright, Theon. *The Big Nail: The Story of the Cook-Peary Feud.* New York: Day, 1970.

CHAPTER

20

Bermuda Triangle Mystery

The Bermuda Triangle mystery, historically speaking, began with Christopher Columbus. When Columbus sailed through the Sargasso Sea and Bermuda Triangle area on his voyage to San Salvador Island in the Bahamas on September 15, 1492, he reported what he called "a remarkable bolt of fire" falling into the sea. His logs contain descriptions of the weed-filled sea, the appearance of a strange light at sea, and the irregular behavior of the ships' compasses. The crew was terrified by these strange events and by their distance from land and home. Columbus was able to reassure his rebellious men, avoid a mutiny, and continue his voyage. Thus the Bermuda Triangle started to take on an air of mystery from its first-known navigator and his superstitious crew almost five centuries ago. It has been theorized that the "great flame of fire" noted by Columbus was a meteor.

The Bermuda Triangle, somewhat narrowly defined, is a triangular region encompassing about 14,000 square miles in the Atlantic Ocean, with the corners of the triangle at the southern Virginia coast, the Bermuda Islands, and the Florida Keys, although the numerous tragedies associated with it reach out to a larger area. Between 1945 and 1972, U.S. Navy and Coast Guard records substantiate the disappearance without a trace of more than 100 vessels and aircraft and more than 1,000 seamen and airmen in this region. In most cases no distress signals were received.

This is the sinister no man's land, surrounded by the popular vacation areas of Bermuda, the Bahamas, and Florida, patrolled by the U.S. government, heavily traveled day and night, and not isolated by any stretch of the imagination.

Early records reveal that four American naval vessels vanished without explanation between 1781 and 1812. In 1840 the *Rosalie,* a large French ship, was found deserted near Nassau, with sails set and carrying a valuable cargo. Mid-nineteenth-century disappearances included the *Bella*, a mystery derelict, found in 1854; the *Lotta*, a Swedish bark that vanished near Haiti in 1866; and the *Viego*, a Spanish merchantman, two years later.

The Bermuda Triangle continued to claim numerous victims in succeeding years. One of the most famous was the *Atalanta* in 1880. She left Bermuda in January headed for England with a crew of three hundred cadets and officers and was never seen again. A massive search was undertaken by a large fleet of ships, but no trace was ever found. Four years later, the *Miramon,* an Italian schooner bound for New Orleans, was lost, and the *Freya*, a German bark, was found abandoned.

A celebrated sailor who had gained renown by sailing alone around the world, Joshua Slocum vanished along with his ship *Spray* in 1909 while crossing the triangle.

Airplanes were due to share in the disasters. On December 5, 1945, five U.S. torpedo bombers left Fort Lauderdale, Florida, on a routine two-hour mission. Fourteen men were aboard the five planes. Shortly after the flight began, the commander reported that he could not see land and believed that he was off course. The five planes were never heard of again. A Martin flying boat was sent out to investigate, and it, too, vanished. In March 1950 an American Globemaster disappeared north of the Triangle; a Jamaica-bound British York transport in February 1953 sent out an SOS, and a search was begun, all to no avail; in 1954 a Navy Super Constellation was lost; two years later a Marine Sky Raider and a Navy patrol bomber with a crew of ten also disappeared. In 1962 an Air Force KB50 left Langley Air Force Base in Virginia headed for the Azores with a crew of nine, and disappeared without a trace.

The U.S. Navy has lost two nuclear submarines, the *Thresher* in 1963 and the *Scorpion* in 1968. Their final voyages ended near the triangle.

The disappearances of ships and planes nearly all had common features: favorable weather conditions, no mechanical problems, and routine radio reports. Intensive searches usually turned up no debris and no bodies.

Supernatural explanations have been offered for the Bermuda Triangle phenomena. In the old days there were tales of sea monsters, giant octopi, squid, and devils waiting to overcome galleons and their crews. Some superstitious folk have believed that demons and devils guided ships to their destruction and, in the modern world, sent airplanes to their doom.

The fact that often not even a remnant was found of missing ships and aircraft has lent weight to another theory: interference by beings from outer space, creatures of superior intelligence with sinister intentions.

Such far-fetched views are rejected by the Coast Guard, which holds that the many Bermuda Triangle happenings are merely coincidence and that "there is nothing mysterious about disappearances in this particular section of the ocean. Weather conditions, equipment failure, and human error, not something from the supernatural, are what have caused these tragedies."

Aviation and marine experts have continued to seek logical explanations for the Bermuda Triangle mystery. They have ruled out sudden tropical storms as the answer, but suspect atmospheric aberrations and electromagnetic gravitational disturbances may be responsible. Some scientists suggest that seaquakes, causing sudden shiftings of the ocean floor and waves up to 200 feet high that can swallow smaller vessels, may be the cause. Naval and other experts do not accept underwater volcanic activity and seaquakes as reasonable causes of ship losses, but concede that freak seas may be responsible. Powerful ocean currents and waterspouts may be factors. Aircraft may be affected by extreme air turbulence.

Extensive investigations by the Navy have concluded, officially, that the disappearances were strange coincidences, and no twilight zone exists where supernatural forces destroy ships and planes. Nevertheless, they concede that the number of disasters in the triangle area exceeds that of more heavily traveled sea and sky areas.

By a curious coincidence, an area off the coast of Japan, known as the Devil's Sea, also has a high rate of disappearances. Between 1950 and 1954, at least nine ships vanished in the Devil's Sea. These were not small fishing boats but large freighters with powerful engines and radios. An SOS was received from only one of the ships. Thorough searches found a few small pieces of debris, but no survivors, bodies, or lifeboats. Shaken by these events, the Japanese government ordered a survey of the area. Nothing was discovered except a new volcano, an eruption that sank one of the research ships. The government then officially proclaimed the area to be a danger zone.

The Bermuda Triangle has been much publicized in books, magazine articles, and radio and television shows. A comprehensive study of its history and analyses of the most famous disappearances appear in Lawrence David Kusche's *The Bermuda Triangle Mystery—Solved.* In addition to cases previously mentioned, Kusche reviews the following:

Mary Celeste (December 1872), a 103-foot-long brigantine of 282 tons was found abandoned at sea by another ship, the *Dei Gratia*, and towed into Gibraltar. The crew of eight plus the captain and his family were missing and never accounted for.

Cyclops (March 1918), a 19,600-ton Navy collier, sailed from Barbados, West Indies, with a crew of 309 men and a cargo of manganese ore. The destination was Norfolk, Virginia, but the ship failed to arrive and a massive search proved fruitless.

Carol A. Deering (January 1921), a five-masted schooner, was discovered grounded on Diamond Shoals with all sails set. The only living things aboard were two cats. A meal was still on the stove. The crew was never found.

Baifuku Maru (April 1925), a Japanese steamship, was headed for Hamburg, Germany, with a cargo of wheat. Caught by heavy weather, an SOS message was sent out and picked up by a White Star liner. Nothing has ever been seen of the freighter and its crew since that fateful morning.

Proteus and *Nereus* (1941). The *Proteus* left St. Thomas, Virgin Islands, on November 23, 1941, headed for Portland, Maine. It never arrived and no trace was ever found. A sister ship, the *Nereus*, following the same route, left St. Thomas on December 10, 1941, and also was lost without a trace.

Rubicon (October 1944), a Cuban cargo ship, was discovered by a Navy blimp off the coast of Florida. The only living creature aboard was a dog. The ship was in excellent condition except for a missing lifeboat and a broken hawser.

Star Tiger (January 1948), a large airliner operated by British South American Airways, was headed for Bermuda. The weather was excellent when the plane disappeared; an unsuccessful search found no oil slicks, bodies, or debris. An investigation by the Ministry of Civil Aviation concluded that the "fate of the *Star Tiger* must remain an unsolved mystery."

Star Ariel (January 1949), a sister airship of the *Star Tiger*, disappeared on a flight between Bermuda and Jamaica; the four-engine plane took off with seven crewmen and thirteen passengers. After a routine flight report from the captain, the *Star Ariel* was never heard from or seen again.

Globemaster (March 1950), an American plane with fifty-three persons on board, disappeared north of the triangle on a flight to Ireland. Apparently the plane was blown to pieces by a terrific explosion.

Sandra (June 1950), a 350-foot freighter, radio-equipped and loaded with 300 tons of insecticides, sailed from Savannah, Georgia, en route to Venezuela. After passing Jacksonville and St. Augustine, Florida, the ship vanished without a trace in a calm sea, with eleven men on board.

A British York Transport (February 1953), with thirty-nine persons, vanished north of the triangle on a flight to Jamaica. No trace was ever found.

A Navy Super Constellation (October 1954), with forty-two persons aboard, disappeared north of the triangle. No emergency message was received from its two powerful radio transmitters. A search by hundreds of planes and ships failed to find a trace.

Southern Districts (December 1954), a connected Navy LST, disappeared in the Straits of Florida with its crew of twenty-two.

A Navy Patrol Bomber (November 1956) vanished near Bermuda without sending a radio call. No debris was found.

Revenoc (January 1958), a sailboat equipped for ocean racing, vanished with its crew of four on a 150-mile trip from Key West to Miami.

An Air Force KB-50 (January 1962) disappeared with nine persons aboard while flying from Langley Air Force Base in Virginia to the Azores. No trace of it was ever found.

A twin-engine Piper Apache (April 1962) was lost while approaching Nassau. The pilot was in sight of his destination when he reported being lost.

Marine Sulphur Queen (February 1963), a 500-foot tank vessel, sailing from Beaumont, Texas, to Norfolk, Virginia, was reported missing when she failed to arrive in Norfolk as scheduled. A few life jackets and a little debris were found, but nothing more.

Sno' Boy (July 1963), a sixty-three-foot charter fishing boat with forty occupants, vanished in good weather on the eighty-mile voyage from Kingston, Jamaica, to the Pedro Banks.

An Air Force Flying Boxcar (June 1965), with ten airmen aboard, mysteriously disappeared when only forty-five minutes from its Bahamas destination.

Witchcraft (December 1967), a cabin cruiser, called for help from the Coast Guard while a mile out to sea from Miami Beach. Response was immediate, but no boat, no men, and no life jackets were found. The boat had vanished while within sight of the city's lights.

Elizabeth (April 1971), a connected LST, left Fort Lauderdale, Florida, with a cargo of scrap paper bound for Venezuela. She was reported passing the Bahamas but was never heard from again.

El Caribe (October 1971), the largest vessel in the Dominican merchant marine, was bound from Colombia to the Dominican Republic. It reported en route but was never seen or heard from thereafter.

Norse Variant and *Anita* (March 1979), two Norwegian freighters carrying coal from Norfolk, Virginia, to Hamburg, Germany, foundered in a storm southeast of Cape May, New Jersey. One member of the *Norse Variant* crew survived and was rescued from a raft several days later. He reported that a hatch cover had been ripped off in the storm and two cargo holds had been flooded, causing the ship to sink within a few minutes.

The foregoing were among the numerous lost ships and planes described by Kusche, all associated in some way with the Bermuda Triangle. Kusche expresses skepticism about any simple solution, stating, after examining all the evidence, that "There is no theory that solves the mystery." The findings of his research are noted under ten headings. Summarized, they are: if sufficient information is available, logical explanations are found for most of the incidents; the unsolved mishaps are generally those for which information is lacking; disappearances occur in all parts of the ocean, not just the Bermuda Triangle; some of the vessels reputedly lost in the Bermuda Triangle actually disappeared elsewhere; the weather was bad when many of the incidents occurred; some were caused by hurricanes; many mishaps occurred in late afternoon or at night, making searches more difficult.

In short, Kusche concludes, "The Legend of the Bermuda Triangle is a manufactured mystery. It began because of careless research and was elaborated upon and perpetuated by writers who either purposely or unknowingly made use of misconceptions, faulty reasoning, and sensationalism. It was repeated so many times that it began to take on the aura of truth."

REFERENCES

Berlitz, Charles I. *Bermuda Triangle.* Garden City, N.Y.: Doubleday, 1974.

Jeffrey, Adi-Kent Thomas. *The Bermuda Triangle.* New York: Warner, 1975.

Kusche, Lawrence David. *The Bermuda Triangle Mystery—Solved.* New York: Harper, 1975.

Snow, Edward. *Supernatural Mysteries.* New York: Dodd, 1974.

Winer, Richard. *The Devil's Triangle.* New York: Bantam Books, 1975.

CHAPTER

21

Loch Ness, Bigfoot, and Other Monsters

LOCH NESS MONSTER

For the past fifty years or more, the Loch Ness Monster has been one of the most famous mysteries of modern times.

The excitement began in May 1933 when a small Scottish local paper carried a report of a gigantic unknown animal being sighted in Loch Ness. A couple driving along the main road that runs along the north shore of the lake reported seeing an enormous creature rolling and plunging in its dark waters. Further sightings followed and by the end of the year the mystery animal was nationally and even internationally famous.

One of the largest and deepest lakes in Scotland, Loch Ness is located in the Great Glen, in the north-central section of the country. The lake is about twenty-five miles long and a mile wide. Loch Ness is part of the Caledonian coral system joining the North Sea with the Atlantic Ocean.

Since the 1933 news story there have been hundreds of sightings reported and numerous photographs taken. Nevertheless, indisputable scientific proof of the monster's existence has yet to be found. The most complete roundup of newspaper stories and photographs is contained in Ronald Binns's *The Loch Ness Mystery Solved.* In his work, Binns reveals that the photographs really show seals, otters, giant eels, a huge fish, floating logs, pieces of driftwood, roe deer, crested grebes, cormorants—anything except a mysterious monster.

In contrast, persons who reported sighting what they assumed was the Loch Ness Monster often gave graphic details: long snake-like heads and necks, huge hump-backed creatures cutting through the water, serpentine bodies, whale-like bodies, furry monsters with legs. There have been comparisons with giraffes and hippopotamuses. Photographs fail to validate such descriptions. Binns comments, "The case for the Loch Ness Monster rests overwhelmingly on eye-witness evidence, and this is where the difficulty lies. No-one has been able to film what eye-witnesses claim they see."

After researching extensively over a period of years, reading the voluminous literature relating to the monster, examining the many photographs, and spending much time on the lake, Binns comes to the following conclusions: "There will be more dramatic snapshots, more amazing eye-witness testimony ... after fifty years one conclusion can reasonably be drawn. There is no scientific evidence whatsoever of monsters in Loch Ness, and a handful of individuals will go on seeing them there."

A lengthy article in the *Skeptical Inquirer*, winter 1984-1985 issue, concludes:

> We have shown that continuous sonar monitoring for seven weeks to a depth of 33 meters in an area where many previous sonar contacts had been reported showed no evidence of anything larger than a 1-meter fish. The circumstances under which previous expeditions had obtained sonar and photographic evidence in support of the existence of the Loch Ness Monster could not withstand scrutiny. The evidence itself revealed discrepancies. This is especially true of the Academy's flipper photographs, the published versions of which differ from the original computer-enhanced photographs. Careless deployment of equipment and overzealous interpretation of the data account for much of the so-called evidence.
>
> While it is not possible to prove definitely that a monster does not exist, the evidence so far advanced strongly suggests that the Loch Ness Monster is nothing more than a longlived and extremely entertaining legend.

BIGFOOT, ABOMINABLE SNOWMAN, AND OTHER APE-MEN

Also celebrated is another type of monster, a mysterious ape-man of the Himalayas. Called the Yeti by the Sherpas, it is also known as the Abominable Snowman, Bigfoot, Man of the Mountains, and Sasquatch.

Stories about the Yeti are legion among the Nepalese and especially among the Sherpas who live in the high mountains. In the imaginations of many scientists, zoologists, and mountaineers, the Yeti is viewed as the missing link between ape and man. It is described as between seven and nine feet tall, wide and ugly of face, covered by long hair, strong enough to wring the necks of full-grown cattle, and fast enough to outrun man and dog. It is said to have attacked young virgins and battered men to death with a club.

The most persuasive evidence of the existence of the Yeti is its giant footprints, from which comes its nickname, "Bigfoot." Lord Hunt, leader of the expedition that first conquered Mount Everest, has photographs taken in 1978 showing huge footprints fourteen inches long and seven inches wide. The weight of the creature was demonstrated by the fact that the steps had broken through a heavy crust of ice, in contrast to men in the expedition who could walk through the snow without making any impression.

A zoologist from London made an exhaustive analysis of footprints brought back by a 1951 expedition and concluded, "All the evidence suggests that the so-called snowman is a very huge, heavily built bipedal primate, most probably of a similar type to the fossil *Gigantopithecus*."

A strange turn in the saga of Bigfoot or Sasquatch is its invasion of the Pacific Northwest. In the wild country on either side of the Cascade Mountains, which run down the Pacific coast of North America from Canada through Washington and Oregon, reported sightings of Bigfoot or Sasquatch now run into hundreds. Dense forests blanket the country along the Columbia River.

Publicity about Bigfoot in the northwestern United States has inspired more than 2,000 accounts of finding footprints or sightings. Furthermore, the creature seems to be wandering far afield. There have been reports from Florida, Tennessee, Michigan, Alabama, North Carolina, Iowa, and Arizona.

OTHER MONSTERS

Canada has three monsters of the deep to rival Scotland's legend of Loch Ness:

1. *Champ*, the monster of Lake Champlain, described as having a long neck and a rather serpentine body up to fifty feet in length;

2. *Ogopogo* in southern British Columbia, estimated at seventy to seventy-five feet in length, with a long flattened head like a snake and two ears standing up from the head;

3. *Manipogo*, a creature of Canada's lakes north of Winnipeg, is said to resemble a large black snake. All are subjects of a number of indistinct photographs, and various indecisive sightings have been reported, but otherwise the Canadian monsters are as elusive as the better-known one in Loch Ness.

Other lake monsters exist in the lore of Sweden, Ireland, New Zealand, Africa, Australia, and Iceland.

Illinois has a famous monster of Indian origin, the Piasa bird. In the language of the Illini Indians, *Piasa* translates as "the bird that devours men."

On the face of a perpendicular bluff overlooking the Mississippi River near Alton, Illinois, is carved the figure of an enormous bird with its wings extended. Father Marquette, who saw the carving in 1673, saw not one but two figures and described them in his journal: "As we descended the river we saw high rocks with hideous monsters painted on them, and upon which the bravest Indians dare not look. They are as large as a calf, with head and horns like a goat; their eyes red; beard like a tiger; and a face like a man's. Their tails are so long that they pass over their heads and between their fore legs, under their belly, and ending like a fish's tail. They are painted red, green, and black."

The Alton *Evening Telegraph* of September 28, 1836, gave what is perhaps the first published account of the story of the Piasa bird.

> The legend of the Piasa or Thunder Bird, as passed on by the Illinois Indians, is as follows: Many thousand moons before the arrival of the palefaces, there existed in this land of green prairies a bird of such dimensions that he could easily carry off in his talons a full grown deer. Having once obtained a taste of human flesh from that time he would prey upon nothing else. He was as artful as he was powerful, would dart suddenly and unexpectedly upon an Indian, bear him off into the caves of the bluff, and devour him. Hundreds of warriors attempted for years to destroy him, but without success. Whole villages were nearly depopulated, and consternation spread throughout the tribes of the Illini. At length, Ouatoga, a chief, whose fame as a warrior extended even beyond the great lakes, separated himself from the rest of his tribe, fasted in solitude for the space of a full moon, and prayed to the great spirit, the master of life, that he would protect his children from the Piasa. On the last night of the fast, the great spirit appeared to Ouatoga in a dream, and directed him to select twenty of his warriors, each armed with a

> bow and poisoned arrow, and conceal themselves in a designated spot. Near the place of their concealment, another warrior was to stand in open view, as a victim for the Piasa, which they must shoot the instant that it pounced upon its prey. When the chief awoke in the morning, he thanked the great spirit, and returned to his tribe, and told them of his dream. The warriors were quickly selected and placed in ambush, as directed. Ouatoga offered himself as the victim. Placing himself in open view of the bluff, he soon saw the Piasa perched on the bluff eyeing his prey. Ouatoga drew up his manly form to its utmost height, and planting his feet firmly upon the earth, began to chant the death song of a warrior. A moment after, the Piasa rose into the air, and, swift as a thunderbolt, darted down upon the chief. Scarcely had he reached his victim, when every bow was sprung and every arrow sent to the feather, into his body. The Piasa uttered a wild, fearful scream, that resounded far over the opposite side of the river, fell into the river, and expired. Ouatoga was safe. In memory of this event, the image of the Piasa was engraved on the bluff.

Other legends prevailing among the American Indians concerned water monsters, including water serpents, water bears, mermaids, mermen, and the Great Fish. Water monsters were frequently painted or carved above pools and streams. Water monsters were common symbols of evil in the cosmogony of the Algonquian and Siouan Indians of the Mississippi Valley.

In his fascinating book, *Bigfoot, the Yeti and Sasquatch in Myth and Reality*, John Napier concludes that "stories of monsters are universals of time and space. The principal strongholds of contemporary monster myths are the mountains of the northern hemisphere.... In North America the coastal ranges of British Columbia and the north-west United States, the Cascade Mountains, the mountains of Guatemala in Central America, the Colombian Massif and the Guyana Highlands in South America." All "are splendid breeding-grounds for monster myths." Napier points out further that monsters usually come from remote areas: "inaccessible mountains, impenetrable forests, remote Pacific islands, the depths of loch or ocean."

African gorillas are a prime source of early monster tales, which were usually complete misconceptions of the nature of the beast. Gorillas were the inspiration for the Tarzan stories and for the movie *King Kong*. As conceived in myths, gorillas are aggressive, sexually rapacious, man-like creatures of Herculean strength—quite unlike the mild-mannered, man-fearing vegetarians they really are.

Sailors and fishermen around the world have circulated tales of giant monsters lurking beneath the waves. One such sea serpent was reportedly seen off the east coast of Brazil. It was described by one witness as "a long serpentine creature; it has a series of humps; its head is rather like a horse's; its color is dark on top and light below; it moves by undulations up and down; it appears during the summer months."

As far back as the fourth century B.C., Aristotle wrote of large serpents off Libya, capable of devouring oxen.

In Norwegian literature, stories abound about the Kraken, the largest of the giant sea monsters. The Kraken when full grown was more than a mile long, with a lumpy spine and tentacles floating like strands of seaweed.

Every culture has had its traditions of monsters and animals of hideous shapes that are usually highly destructive. The ancient Greeks and Romans told tales of centaurs and hydras; the Moors and Egyptians told stories of man-eating and other fearsome creatures. The English have transmitted the legend of the winged dragon vanquished by St. George. Folklorists and historians have suggested the possibility that mastodons, mammoths, saber-toothed tigers, or other extinct beasts may have been the original inspiration for such tales.

In his discussion of monsters from around the world, John Napier notes that "Mythical creatures *have* turned up in real life." The unicorn, for instance, may have been seen in the Arabian oryx, and the dragon in the Komodo lizard.

REFERENCES

Binns, Ronald. *The Loch Ness Mystery Solved.* Buffalo, N.Y.: Prometheus Books, 1984.

Napier, John. *Bigfoot, the Yeti and Sasquatch in Myth and Reality.* New York: Dutton, 1973.

Razdan, Rikki, and Alan Kielar. "Sonar and Photographic Searches for the Loch Ness Monster: A Reassessment." *The Skeptical Inquirer* 9 (Winter 1984-1985): 147-57.

CHAPTER

22

Astrology

Astrology aspires to be an exact science. Astrologists engage in extensive calculations, including the precise time of birth, the sign of the zodiac prevailing when a person comes into the world, the positions of the sun and the stars, aspects of the various planets, the ascendants, the "houses" in which the heavenly bodies happen to be at the moment of birth.

According to astrology, man's years are divided into seven parts, governed by the seven planets. The first is infancy, four years, ruled by the moon; then comes childhood, ten years, governed by Mercury; youth, eight years, over which Venus presides; adolescence, twenty years, ruled by the sun, when man attains his full strength and vigor; manhood, fourteen years, under the dominion of Mars, a bad star, when man becomes angry, impatient, and avaricious; old age, twelve years, governed by Jupiter; and finally decrepitude, ruled by Saturn.

Astrology and astronomy developed together in Chaldean Babylonia, beginning in 3000 B.C. or earlier. Astronomy provided a calendar to regulate hunting and sowing activities, while early astrology was concerned with omen predictions. The Babylonians were excessively superstitious. They saw portents in every kind of phenomenon, from simple, everyday happenings to the state of the entrails of specially sacrificed animals. Biblical references to early astrology date from about 700 B.C., with several examples in the Old Testament. By the Middle Ages, astrology had spread throughout the world.

The peak of astrologers' influence was from the fourteenth to the seventeenth centuries, during which time they guided the decisions of monarchs throughout Europe. Court astrologers were key people in royal courts, and kings turned regularly to them for advice. Louis XI (1423-1483), a French king, was known to have consulted an astrologer regularly in his devious plans and complex politics. In reference to another French king, Charles the Wise (1337-1380), Thomas Watson wrote, "for all his wisdom, Charles was a firm believer in astrology, and a state astrologer was one of the honored and salaried officials of his administration. It was this man's sworn duty to tell the King what was going to happen, so that the King *might take measures to keep it from happening.*"

In England, Henry VIII had a professional astrological adviser, Nicholas Kratzer, a Bavarian by birth. A more famous court astrologer was John Dee, adviser to Elizabeth I, who was believed to have exerted a subtle influence on many of the Queen's decisions.

The most celebrated astrologer of the sixteenth century, perhaps of all time, was Nostradamus (1503-1566), adviser to Catherine de Medici. The first edition of his famous *Centuries* appeared between 1555 and 1558. These are prophecies in verse of an involved and mystical nature, some of which are believed to have come to pass, such as Britain's emergence as a great empire and the French Revolution.

Converts to astrology often point to two notable sixteenth-century astronomers to support their cause: Tycho Brahe and Johannes Kepler. Brahe, a Dane, made a detailed study of a new star that appeared in 1572 and linked various astrological forecasts to it. He also made predictions about the weather, the relation of astrology to political events, and the deaths and accessions of kings. He became well known for his interpretations of eclipses and their astrological effects on the ruling houses of Denmark and Norway. Frederick II of Denmark presented Brahe with an island called Hveen off the Danish coast. There he remained for twenty years, engaged in astronomical work and astrological forecasting using a number of imported astronomical instruments. His attitude toward astrology is revealed by his statement that "astrology is not a delusive science when kept within bounds and not abused by ignorant people," indicating a degree of skepticism, though he appears never to have doubted the validity of astrological predictions.

Johannes Kepler joined Tycho Brahe as an assistant at Prague in the last year of Brahe's life. There he formulated the planetary laws for which he is famous. He earned his living, however, by publishing an annual almanac and doing detailed astrological forecasting for Wallenstein, the great Austrian general. Kepler often complained about the onerous duties of preparing almanacs and casting charts. Although there seems no doubt that he believed in astrological forecasting, he criticized professional astrologers for being a vulgar and childish race of dreamers.

The first step in astrological forecasting is to calculate the birth chart, a stylized map of the heavens at the moment of birth. The chart includes the twelve signs of the zodiac, the planets (taken from a book of planetary positions called an ephemeris), and the "houses." The planets include the sun and the moon. The twelve zodiac signs, their dates, and ruling planets are as follows:

Aries, the ram, March 21-April 19	Mars, Pluto
Taurus, the bull, April 20-May 20	Venus
Gemini, the twins, May 21-June 21	Mercury
Cancer, the crab, June 22-July 22	Moon
Leo, the lion, July 23-August 22	Sun
Virgo, the virgin, August 23-September 22	Mercury
Libra, the scales, September 23-October 23	Venus
Scorpio, the scorpion, October 24-November 22	Mars, Pluto
Sagittarius, the archer, November 23-December 21	Jupiter
Capricorn, the goat, December 22-January 20	Saturn
Aquarius, the water-bearer, January 21-February 18	Uranus
Pisces, the fish, February 19-March 20	Neptune, Jupiter

The signs are divisions of the zodiac, each thirty degrees in extent. Another important astrological series is the "houses," numbering twelve. Six of the divisions are above the horizon and six below. In the course of the twenty-four-hour day every celestial object will appear to swing round through all twelve houses, taking an average of two hours to pass through each one.

There are other factors that may be taken into account by the astrologer, such as stars, asteroids, and eclipses.

A horoscope is a diagram of the heavenly bodies showing the relative position of the sun, moon, stars, and planets at a given time. An individual's horoscope is based on the exact time and place of his birth. Each of the twelve signs of the zodiac is believed to be associated with definite aspects of character, temperament, physiology, and aptitude. Knowing the positions of the heavenly bodies at the exact time of a person's birth, astrologers claim to be able to predict his future and advise him about reaching correct decisions.

Beginning in the seventeenth century, astrology began a steady decline. With discoveries of Copernicus, Galileo, Newton, Kepler, and other giants of science, the planets assumed their rightful positions—huge celestial objects, like the earth, circling the sun according to known physical laws. Nevertheless, millions of people continue to believe astrological observations and are faithful readers of syndicated newspaper columns with their daily horoscopes. In response to the popular acceptance of astrology, one commentator notes that millions of people have also believed that the earth was flat.

Serious astrologers, by the way, are scornful of popular newspaper forecasts, pointing out that they are based entirely upon the sign of the zodiac when the person comes into the world, taking no notice of the positions of the stars, aspects of the various planets, the "houses" in which the heavenly bodies happen to be at the moment of birth, and other factors.

A damning indictment of astrology is contained in a penetrating article by Steven Goldberg, a member of the City University of New York faculty. After a thorough analysis of the claims and counterclaims, Goldberg answers the question "Is Astrology Science?" in a 1979 issue of the *Humanist*. The conclusion: "Clearly there is not a scintilla of evidence of any kind in favor of astrology and there is an enormous amount of indirect theoretical evidence against astrology (that is, the astronomical evidence that casts the greatest doubt on the possibility that distant stars exert the sorts of influence astrology implies they exert)."

Perhaps the definitive word on the subject comes from a statement by 192 leading scientists, including nineteen Nobel Prize winners. Entitled "Objcctions to Astrology," the statement reads as follows:

> Scientists in a variety of fields have become concerned about the increased acceptance of astrology in many parts of the world. We, the undersigned—astronomers, astrophysicists, and scientists in other fields—wish to caution the public against the unquestioning acceptance of the predictions and advice given privately and publicly by astrologers. Those who wish to believe in astrology should realize that there is no scientific foundation for its tenets.
>
> In ancient times people believed in predictions and advice of astrologers because astrology was part and parcel of their magical world view. They looked upon celestial objects as abodes or omens of the Gods and thus intimately connected with events here on earth; they had no concept of the vast distances from the earth to the planets and stars. Now that these distances can and have been calculated, we can see how infinitesimally small are the gravitational and other effects produced by the distant planets and the far more distant stars. It is simply a mistake to imagine that the forces exerted by stars and planets at the

moment of birth can in any way shape our futures. Neither is it true that the position of distant heavenly bodies makes certain days or periods more favorable to particular kinds of action, or that the sign under which one was born determines one's compatibility or incompatibility with other people.

Why do people believe in astrology? In these uncertain times many long for the comfort of having guidance in making decisions. They would like to believe in a destiny predetermined by astral forces beyond their control. However, we must all face the world, and we must realize that our futures lie in ourselves, and not in the stars.

One would imagine, in this day of widespread enlightenment and education, that it would be unnecessary to debunk beliefs based on magic and superstititon. Yet, acceptance of astrology pervades modern society. We are especially disturbed by the continued uncritical dissemination of astrological charts, forecasts, and horoscopes by the media and by otherwise reputable newspapers, magazines, and book publishers. This can only contribute to the growth of irrationalism and obscurantism. We believe that the time has come to challenge directly and forcefully the pretentious claims of astrological charlatans.

It should be apparent that those individuals who continue to have faith in astrology do so in spite of the fact that there is no verified scientific basis for their beliefs, and indeed there is strong evidence for the contrary.

REFERENCES

Eysenck, H. J. *Astrology*. New York: St. Martin's Press, 1982.

Goldberg, Steven. "Is Astrology Science?" *The Humanist* (March-April 1979): 9-16.

Jerome, Lawrence. *Astrology Disproved*. New York: Prometheus Books, 1977.

Kurtz, Paul, and Andrew Fraknoi. "Scientific Tests of Astrology Do Not Support Its Claims." *Skeptical Inquirer* 9 (Spring 1985): 210-11.

Lyndoe, Edward. *Astrology for Everyone.* New York: Dutton, 1960.

MacNeice, Louis. *Astrology.* Garden City, N.Y.: Doubleday, 1964.

Naylor, P. H. *Astrology.* North Hollywood, Calif.: Wilshire Book Co., 1974.

CHAPTER

23

Alchemy

The science of chemistry began in the Middle Ages as alchemy, a mixture of black magic, a limited amount of scientific knowledge, and a great deal of superstition. The chief goal of alchemists was to find a mythical "philosopher's stone" that could be used to transform the base metals, such as iron and lead, into gold. They were also trying to find or to produce an elixir of life that would make them live forever.

Alchemy began about A.D. 100. The idea of transmuting base metals into the nobler ones of gold and silver probably originated among the Alexandrian Greeks in the early centuries of the Christian era. The writings of the classical Greek philosophers, especially Aristotle, were an inspiration. Aristotle maintained that all substances originated in some way from four basic elements—earth, air, fire, and water. The aim of the alchemists was to find the "quintessence," the fifth element, which could control the changing of one substance into another.

The concept of transmutation was influenced by ideas from the East. The earliest Greek alchemical writings contain numerous references to oriental authorities and traditions. Radical changes followed the conquest of Alexandria by the Moslems in 640. The Arabs were captivated by Greek science, especially alchemy, and translated the literature on the subject into Syrian and Arabic.

One of the earliest and greatest of Arabian alchemists was Jabir, who flourished in the eighth century. Jabir taught that solid matter contained a metallic substance called mercury and a flammable one called sulfur. By proper manipulation of these substances, gold could be formed out of base metals. An eighth-century Spanish alchemist and metallurgist, Geber, discovered nitric acid. In the ninth century the Arabian alchemist Al-Razi described plaster of paris and other new materials, and a short time later the physician Avicenna summarized Arabian advances in alchemy, but expressed doubt that transmutation was possible.

Thereafter, Arabic alchemy made no important progress, while from the twelfth century on there was renewed interest in Europe. The most significant discoveries during that period were arsenic by Albertus Magnus (c. 1200-1280); gunpowder by a German alchemist, Berthold Swartz, in the thirteenth century; and mineral acids by a Spanish alchemist, Geber (in the eighth century). An influential figure was a Swiss physician, Paracelsus (c. 1493-1541), who denounced the alchemists' preoccupation with the search for gold and declared that their main purpose should be a search for new medicines. A German alchemist, Libarius, in 1597 was the first to describe the formation of hydrochloric acid.

However, as the science historian Isaac Asimov pointed out, "the search for gold and for the elixir of life completely obscured these occasional real and worthy advances of honest scholars. Alchemy became a happy hunting ground for ignorant jargon-mouthing frauds and rascals. These found dupes among gold-seeking nobility and royalty and, occasionally, gallows for themselves."

Charles MacKay, in his *Extraordinary Popular Delusions*, observed that for more than a thousand years millions of people believed in the art of alchemy. He offers biographical sketches of the most celebrated, or in some cases most notorious, characters who played roles during that time, beginning with Geber in the eighth century to Cagliostro in the eighteenth century. Among the memorable names are the following:

Alfarabi, a tenth-century figure who was one of the most learned men of his age. He spent his life traveling from country to country, seeking the elixir of life and the secret of goldmaking.

Artephius, twelfth-century author of two famous treatises, one on the philosopher's stone and the other on the art of prolonging human life.

Arnold de Villeneuve, born in 1245, reputed to have been the most able physician of his time and thought to have made much gold from lead and copper.

Roger Bacon, born in 1214, a firm believer in the philosopher's stone who spent much time searching for it.

Jacob Boehme, born in 1575, inventor of a strange doctrine combining alchemy and religion, and author of *Metallurgia* and other works.

Robert Boyle (1627-1691), famous for his *The Sceptical Chymist*, which disproved current ideas about the composition of matter. He was one of the founders of the Royal Society of London.

Other names that stand out in the history of alchemy are Vincent de Beauvais, who described in his book *Speculum Majus* (c.1256), the doctrine of mineral formation from vapors; Agricola, author of *De Re Metallica*, translated by Herbert Hoover and his wife; Cagliostro, born about 1743, called the "arch-quack of his age" by MacKay, and one of the most conspicuous characters of Europe; Jan Baptista van Belmont (1577-1644), who studied carbon dioxide, originated the word *gas*, and made the first biochemical experiments; Johann Rudolf Glauber (1604-1668), who discovered sodium sulfate; and Hennig Brand, who discovered phosphorus in 1669. The name of Sir Isaac Newton (1642-1727) should probably be added, for Newton occupied himself for years with attempts to form gold by transmutation.

After the King of Prussia ordered the hanging of the famous alchemist Caetano in 1709, itinerant gold makers went out of style. French chemist Antoine Laurent Lavoisier (1743-1794) laid a firm foundation for modern chemistry with his writings, after which alchemical ideas passed out of respectable science.

MacKay correctly observes that "The useful science of chemistry is not a little indebted to its spurious brother of alchemy. Many valuable discoveries have been made in that search for the impossible, which might have been hidden for centuries yet to come." Another commentator, William F. Kiefler, editor of the *Journal of Chemical Education*, adds, "Although their efforts were largely misdirected, the alchemists contributed much useful information as a result of their experiments. Special kinds of equipment, such as the test tube, the closed crucible, and the retort, still used in laboratories today, were devised by the alchemists."

REFERENCES

American Peoples Encyclopedia. Vol. 1, p. 308. New York: Grolier, 1969.

Burland, Cottie Arthur. *The Arts of the Alchemists*. New York: Macmillan, 1968.

Federmann, Reinhard. *The Royal Art of Alchemy*. Philadelphia: Chilton, 1969.

Hutin, Serge. *A History of Alchemy.* New York: Walker, 1962.

Read, John. *Prelude to Chemistry.* London: Bell, 1936.

Redgrove, E. Stanley. *Alchemy.* New Hyde Park, N.Y.: University Books, 1969.

CHAPTER

24

Judging Character

Phrenology, Physiognomy, Palmistry, Graphology

Over time, a variety of pseudoscientific methods have been offered for measuring character and personality based on certain physical features. Best known are phrenology, physiognomy, palmistry, and graphology.

Aristotle believed that the head was a kind of cooling tower filled with water. Nearly two centuries ago, phrenology, the so-called science of character reading by the study of prominences on the skull, began to attract wide popular interest and attention in Europe and America. From ancient times the brain was thought to be the seat of the intellect and the emotions. Followers of Galen, the celebrated Greek physician, thought that there was a connection between the shape of the head and character and disposition. Later, in the thirteenth century, Albertus Magnus prepared a detailed description of the distribution of mental and physical faculties within the head.

Modern phrenology began in 1796, introduced by Franz Joseph Gall, a German physician. Gall and his disciples argued that human personality consisted of a number of independent, inborn mental "faculties," each of which was localized in a part of the brain. The larger the size of each region, the stronger the faculty. A person's character would be revealed by an examination of skull bumps. According to phrenologists, heads indicate genius, potential criminals, morons, and maniacs, while facial characteristics—the profile, size, and shape of the forehead, hair, ears, and eyes—indicate personality.

Gall's pupil Johann Gasper Spurzheim coined the word *phrenology,* toured Great Britain and the United States, and spoke to crowded lecture halls. One of his influential converts was George Combe, a lawyer who became a leading exponent in England. In 1838-1839, Combe lectured in the United States, where he was elected a member of the American Philosophical Society and the National Academy of Sciences. When he lectured at Yale University, most of the faculty were in his audience. In Washington, the surgeon general of the United States was secretary of the local phrenological society. Of 144 members in Boston, one-third were medical doctors.

Eminent converts helped to spread the doctrine. Henry Ward Beecher declared that "a practical knowledge of the human mind as is given in phrenology" is the best preparation for a Christian. The noted educator Horace Mann expressed indebtedness to phrenology for many of his educational ideas. Arthur Conan Doyle was both a spiritualist and a believer in phrenology. When Sherlock Holmes concluded from the large size of a hat that the wearer was "highly intellectual," he was following a phrenological principle. Sir Isaac Pitman's large hat size was assumed to have contributed to his inventiveness in shorthand and phonetic systems. Walt Whitman took so much pride in the fact that phrenology showed all his faculties to be well developed, he had a chart of his own head bumps published on five occasions. *Leaves of Grass* is filled with phrenological terms. Alfred Russel Wallace, co-discoverer with Charles Darwin of the natural selection mechanism for organic evolution, wrote in 1899, "In the coming century phrenology will assuredly gain general acceptance. It will prove itself to be the true science of mind. Its practical uses in education, in self-discipline, in the reformatory treatment of criminals, and in the remedial treatment of the insane will give it one of the highest places in the hierarchy of the sciences."

Phrenological charts developed by Combe and other practitioners show the human skull divided into thirty-seven areas or traits of character. They are labeled with such sentiments as self-esteem, love of approbation, cautiousness, benevolence, veneration, firmness, conscientiousness, hope, wonder, and other attributes.

Dr. Allen MacLane Hamilton, in his *Recollections of an Alienist*, relates an amusing instance of the misapplication of phrenology:

> In an Italian murder trial ... much testimony had been given by an anthropologist and craniologist in regard to the peculiar shortness of the head of the defendant. So improbable was his testimony that on my way downtown the next morning I stopped at the shop of Dunlap, the hatter, and procured a number of *conformateur* tracings, which I gave to the District Attorney,

> whose first words to the witness who resumed his place on the stand were, "Doctor, *extreme length* of the head is also a stigma of insanity, is it not?" To which the witness at once assented. "Now, Doctor, I will show you some tracings and get you to give your opinion on their meaning." The witness at once admitted that they looked as if they might have been made from the heads of insane people—in fact he was quite sure that all these people were insane to some degree or words to that effect. "Well," said his merciless cross-examiner, "Would it surprise you to know that the first series were taken from the heads of William H. Vanderbilt and his sons, and the *last is from the head of his Honor, the judge, who presides in this case.*" The explosion that followed was allowed to pass without reprimand from the Court.

The tremendous vogue for and popular interest in phrenology continued throughout the nineteenth century. It claimed to be able not only to diagnose faculties, to analyze crime, and to provide a clue to mental disorders, but to guide education and every aspect of the study of the science of man. Modern research on the brain and the study of psychology, together with ridicule by scientists, has demonstrated that phrenology is without merit as a science. The intelligence of a person and the size of his head have no direct relationship. Some of the largest brains known to science have belonged to idiots. As summarized by the *Encyclopaedia Britannica*, "psychology, physiology, and experience alike contribute to discredit the practical working of the system and to show how worthless the so-called diagnoses of character really are."

Physiognomy, the art of judging character from facial features, is closely related to phrenology. It may have begun with ancient Greeks. A treatise ascribed to Aristotle examined the proportions and development of the face, head, neck, limbs, and torso and what they meant in relation to character and temperament. The basic principle of Aristotelian physiognomy was animal resemblance, e.g., a man with a thick bulbous nose like a pig's has the pig's lack of sensibility; a sharp-tipped nose like a dog's indicates the snapping, snarling disposition of a canine; a large, round, blunt nose like a lion's is a sign of magnanimity; a slender, hooked nose like an eagle's indicates a noble but grasping nature.

The connection between facial features and criminal behavior was explored by an Italian occultist, Cesare Lombroso (1836-1909). His findings are regarded as having little scientific merit. Later, Earnest A. Hooton of the Harvard University anthropology faculty undertook research on the same subject. Hooton found many correlations with certain types of criminality. Robbers tended to have heavy beards, diffused pigment in the iris, attached earlobes, and various other body traits. His conclusions have been only partially accepted by other researchers in the same field.

Gardner's *In the Name of Science* points out that persistent mental attitudes may leave an impression on facial features. For example, a person who has felt gloomy for many years is likely to look gloomy, or a person who laughs a great deal may develop mirth wrinkles about the eyes. Sociologists express skepticism, however, about such folk beliefs as red-haired persons having fiery tempers or a jutting chin being a sign of determination.

An excellent statement on physiognomy is contained in John Brophy's *The Human Face Reconsidered* (1962):

> We do read character in faces, and sometimes accurately and with penetration, but this is when we either know the owner of the face intimately, or possess other evidence about him from words spoken or written, from his actions, and reactions, with the mutable expressions of the face which accompany them, from the voice, from the total physical presence.

Brophy goes on to list a dozen popular beliefs about physiognomy:

1. Large, well-spaced eyes: candor and honesty.
2. Small eyes, set close together: cunning and treachery.
3. Eyes that avoid your face: unreliability or guilt.
4. Eyes with deep upper lids: secrecy, suspiciousness.
5. Long nose: inquisitiveness.
6. Nose with down-turning tip: avarice.
7. Wide mouth: generosity, affability.
8. Thick lips: sensuality.
9. Thin lips: envy.
10. Narrow mouth: primness, self-righteousness.
11. Square or jutting chin: determination, heroism.
12. Receding chin: weakness, indecision.

Brophy reports that he could find no evidence for accepting any of these widely accepted popular beliefs. He believes that the shape of the nose, ears, lips, the color of hair and eyes, the texture of the skin, and other facial features are unrelated to mental traits.

Daniel W. Hering, in his *Foibles and Fallacies of Science*, applies various principles of physiognomy to two famous musicians: Beethoven and Mozart. Beethoven's nose was short and stubby; Mozart's was long and pointed. Beethoven had a shock of curly and unmanageable hair; Mozart's hair was fine and silky. Beethoven's cheeks were pudgy; Mozart's were thin. Beethoven's face was short and broad; Mozart's long and narrow. Mozart's hands were well adapted to playing the piano or for fingering any musical instrument; Beethoven's fingers were flat at the ends and so short that he could not span an octave. Both artists were short, about five feet six in height, but Beethoven's figure was stocky, while Mozart was slender. Which characteristics marked the musician?

Like phrenology and physiognomy, palmistry is another attempt to relate shapes and human destiny. In the case of palmistry, the lines, crevices, bumps, and bulges on the hand are claimed to be the key to personality. Palmists classify hands in four types: *practical*, with square palms and short fingers; *intuitive*, with long palms and short fingers; *sensitive*, with long palms and long fingers; and *intellectual*, with square palms and long fingers.

Palmistry is an ancient art. The Hindus were perhaps the first people to practice it, and the Chinese studied the mounts of the hand some 5,000 years before the Christian era. Ancient Greek philosophers, including Aristotle, wrote at length about palmistry. The practice was popular in the Middle Ages, though the Church frowned upon it. In the United States wandering gypsies have largely monopolized palmistry for fortune telling. Needless to say, it has no status in modern science.

Palmistry claims to have the ability to diagnose disease, to judge character, and to foretell the future. There is no scientific basis for any of the claims.

A chart of the hand, for the palmist's purpose, shows the forefinger labeled Jupiter, the second finger Saturn, the ring finger Apollo, and the little finger Mercury. A prominent vertical line running to Saturn was called the Line of Fate. The subject's fortune could be predicted depending upon whether the line was broad and straight, wavy, or double. A variety of other interpretations account for the lines, mounts, and marks of the hand and the general shape of the hand and fingers—all assumed to reveal character and fate.

The concept of using handwriting as a basis for judging character—graphology—appears to have originated in Italy in the early seventeenth century. Further inspiration came in the mid-nineteenth century when the Abbe Jean Hippolyte Michon of Paris developed an elaborate set of "signs" consisting of the shape of loops, form of *T* crosses, position of *i* dots, etc., all of which were related to specific mental traits. Recent research has invalidated most of these conclusions.

Graphology, however, has not been rejected as completely by the experts as have been phrenology, physiognomy, and palmistry. It seems to have a sound psychological basis. For instance, a lazy, easygoing person is likely to be sloppy in writing, while a neat, orderly person will produce a more precise, formal style. Some psychologists suggest that handwriting may be viewed as a form of behavior similar to talking, walking, facial expression, laughing, or shaking hands. Graphologists have analyzed the tendency to lean handwriting from right to left, the size and proportions of writing, the angularity or roundness, the width of margins, and other factors in studying unconscious behavior. An analysis by a graphologist looks at specific letters—their size, shape, slant, and how they are grouped. The individual's signature is also studied. In all, more than 300 aspects of a person's script are examined. The validity of many such criteria as applied to individual personalities is likely to remain a matter of continuing debate.

REFERENCES

Graphology

Jacob, Hans J. *Analysis of Handwriting.* London: Allen & Unwin, 1968.

Olyanova, Nadya. *Handwriting Tells.* Indianapolis, Ind.: Bobbs-Merrill, 1969.

Saudek, Robert. *The Psychology of Handwriting.* London: Allen & Unwin, 1954.

Palmistry

Brockman, Manuel. *The Complete Encyclopedia of Practical Palmistry.* Englewood Cliffs, N.J.: Prentice Hall, 1972.

Hamon, Louis. *Cheiro's Complete Palmistry.* New Hyde Park, N.Y.: University Books, 1968.

Phrenology

Davies, John D. *Phrenology: Fad and Science; A 19th Century Crusade.* New Haven, Conn.: Yale University Press, 1955.

Fowler, Orson S. *Phrenology.* New York: Chelsea House, 1969.

Leek, Sybil. *Phrenology.* New York: Macmillan, 1970.

McCoy, Robert W. "Phrenology and Popular Gullibility." *The Skeptical Inquirer* 9 (1935): 261-68.

Physiognomy

Brophy, John. *The Human Face.* New York: Prentice-Hall, 1946.

Gardner, Martin. *In the Name of Science.* New York: Putnam, 1952.

Mitchell, M. E. *How to Read the Language of the Face.* New York: Macmillan, 1968.

CHAPTER

25

Dreams

In the ancient world, the prevailing belief was that dreams were sent by the gods. They were regarded as divine visitations or prophecies, wanderings of the soul, or even actual events. Prophetic dreams are mentioned in many Middle Eastern and Asian writings, including the Bible. It was not until the middle of the nineteenth century that the general belief in the divine source of dreams began to change. During that period, Alfred Maury made an exhaustive study of dreams and concluded that they resulted from the misinterpretation of sense impressions during sleep. For example, a loud sound during the night might cause the sleeper to dream of a thunderstorm. According to modern theories of dreams, they are extensions of the waking state.

Charles MacKay, in his *Extraordinary Popular Delusions* (first issued in 1841), points out that "the art of interpreting dreams is a relic of the most remote ages.... The records of five thousand years bear abundant testimony to the universal diffusion of the belief that the skillful could read the future in dreams."

MacKay adds, "In many parts of Great Britain, and the continents of Europe and America, there are to be found elderly women in the villages and country-places whose interpretations of dreams are looked upon with as much reverence as if they were oracles. In districts remote from towns it is not uncommon to find the members of a family regularly every morning narrating their dreams at the breakfast-table and becoming happy or miserable for the day according to their interpretation."

In recent years, there has been some resurgence of the belief that dreams come from supernatural sources, God or some gods.

One evidence of the preoccupation with dreams was the great demand at one time for dream books. Such works as *Mother Bridget's Dream-Book and Oracle of Fate* and the *Norwood Gipsy* went through numerous editions and sold in hundreds of thousands of copies.

William Jones, in his *Credulities Past and Present*, confirms MacKay's findings. He notes that "Indications of good or ill luck by dreams are innumerable, and have found believers in the past ages; and we may add, in a milder form of credulity, even at the present time."

The founder of modern psychiatry, Sigmund Freud, was one of the most prolific scientific writers of the twentieth century. A variety of new concepts and psychological contributions emanated from his pen. Freud's own favorite was his earliest major work, *The Interpretation of Dreams*, issued in 1900, which contains nearly all of his fundamental observations and ideas.

Freud found that the analysis of dreams is a fruitful device for probing into inner conflicts and emotions, a field in which he was a pioneer. Before his time, scientists generally regarded dreams as without meaning or purpose. *The Interpretation of Dreams* was the first attempt at a serious scientific study of the phenomenon. Thirty-one years after publication of the book, Freud remarked that "It contains, even according to my present-day judgment, the most valuable of all the discoveries it has been my good fortune to make." According to Freud, "We are justified in asserting that a dream is the disguised fulfillment of a repressed wish." Each dream represents a drama in the inner world. "Dreams are invariably the product of a conflict," states Freud, and "The dream is the guardian of sleep." Its function is to aid rather than disturb sleep, releasing tensions that come from unattainable wishes.

Freud classified mental activity in an individual as occurring on three levels, which he named the Id, the Ego, and the Superego. Of first importance is the Id, the center of primitive instincts and impulses. It reaches into man's animal past, and is animal and sexual in nature. The Ego, as Freud put it, is the mediator "between the reckless claims of the Id and the checks of the outer world." The third element in the mental process is the Superego, which may be broadly defined as conscience.

The dream world, in the Freudian view, is dominated by the unconscious, by the Id. Dreams are important to the psychoanalyst because they lead him into the patient's unconscious. In the unconscious are all the primitive wishes and emotional desires suppressed from conscious life by the Ego and Superego. Always present under the surface, the animal desires force themselves forward in dreams. Even in sleep, however, the Ego and

Superego stand as censors. For that reason, the meanings of dreams are not always clear. Because they are expressed in symbols, they cannot be taken literally, except perhaps in the simple dreams of children. In *The Interpretation of Dreams*, Freud offers numerous examples of dreams that he has psychoanalyzed.

A second famous investigator of dreams was the Swiss psychologist and psychiatrist Carl Jung. For Jung, as for Freud, dreams are a means of exploring otherwise hidden regions of the unconscious mind. Like Freud, he believed that dreams are mainly symbolic and serve as compensation for aspects of the dreamer's personality that are not fulfilled in his waking state—similar to Freud's wish-fulfillment theory. Unlike Freud, who associated dreams with infantile wishes, Jung believed that dreams originate in inborn thought patterns of a racial nature. Jung drew on mythology, comparative religion, and history in interpreting dreams as symbols.

Both Freud and Jung were principally concerned with dreams for psychoanalytical purposes. A new approach was made in 1953 by two University of Chicago scientists, Nathaniel Kleitman and Eugene Aserinsky, when they reported observing the eye movements of sleepers. They found that a series of rapid eye movements (REM) occurred from four to six times during the night. About one-fifth of an adult's typical sleep is REM sleep, much higher in infants and somewhat less in the elderly. When subjects were awakened during REM periods, they usually were able to recall and relate their dreams.

The belief now is that dreaming may be more or less continuous during sleep, but dreams are more likely to be recalled if they occur during REM periods. Further, the indications are that there is an urgent need for REM sleep. If time is lost by being awakened, REM sleep is greatly increased for several nights until the lost time is recovered. It seems to be conclusively demonstrated that everyone normally dreams every night, even though dreams may not be remembered.

As one authority, Calvin S. Hall, of the University of California, points out, "Understanding of dreams has been greatly advanced in the twentieth century by the analytic work of Freud and Jung and the discovery of REM by Kleitman and Aserinsky. No longer are dreams regarded as messages sent by gods or ancestors. Today they are being studied scientifically to learn more about man's nature."

Psychoanalysts hold a variety of opinions as to the value of dream analysis. Many present-day psychiatrists tend to view dreams as attempts to solve problems rather than as the fulfillment of unconscious desires. Experts agree that dreaming is essential to mental and physical well-being.

The popular attitude towards dreams is illustrated by a best-selling book, Alan Davis's *What Your Dreams Mean*. Using the theories of Freud,

Jung, and other psychologists as points of departure, Davis lists actions, objects, characters, and settings that seem most significant when they appear in dreams. The author proceeds to compile a dream dictionary, alphabetically arranged by subjects from *acropolis* to *zoo*. Guided by the dictionary, each individual is advised how to understand his or her dreams. At the same time, it is recommended that everyone keep a dream log, "because dreams are meaningful only in relation to other dreams."

REFERENCES

Bauer, W. W. *Insomnia, Sleep and Dreams.* New York: Simon & Schuster, 1967.

Foulkes, David. *Children's Dreams.* New York: Wiley, 1982.

Freud, Sigmund. *The Interpretation of Dreams.* New York: Macmillan, 1913.

Hadfield, J. A. *Dreams and Nightmares.* London: Penguin, 1954.

Hall, Calvin S. *The Meaning of Dreams.* New York: Harper, 1966.

Hall, Calvin S., and Robert L. Van de Castle. *The Content Analysis of Dreams.* New York: Appleton, 1966.

Jung, Carl G. *The Structure and Dynamics of the Psyche.* London: Routledge, 1953.

Kleitman, Nathaniel. *Sleep and Wakefulness.* Chicago: University of Chicago Press, 1963.

Mackenzie, Norman. *Dreams and Dreaming.* New York: Vanguard, 1965.

CHAPTER

26

Flying Saucers—UFOs

Since biblical times, observations of strange objects moving across the sky at phenomenal speeds have been reported. Popularly known as flying saucers or UFOs, unidentified flying objects, interest in these mysterious phenomena has grown in recent years. Thousands of sightings have been reported in the past forty years. Some UFOs are described as disc-shaped and several hundred feet in diameter, others as oblong or cigar-shaped. Many are reported as traveling at high speeds, making strange changes in direction, and hovering noiselessly. Sightings seen at night are often of colored lights that behave in eccentric ways.

An experience by a businessman, Kenneth Arnold, flying his private plane in 1947 over the Cascade Mountains in search of the wreck of a missing plane, greatly stimulated interest in the subject. Near Mount Rainier, Arnold saw a string of bright metallic objects flying in single file. There were nine: eight discs and one crescent. Their speed was estimated at 1,700 miles per hour and their size as equal to DC-4 airliners. They dodged around the mountain peaks with "flipping, erratic movements." Arnold compared the objects to plates or saucers being skipped over water. Thus was coined the term *flying saucers*.

The Arnold exploit was given wide publicity. Sightings of flying saucers spread like an epidemic all over the United States and elsewhere. In the following years innumerable sightings were reported of flying discs, doughnuts, flaming cigars, fireballs, and other assorted objects.

In January 1948, people in various parts of Kentucky reported seeing a large glowing object in the sky. Air Force Captain Thomas Mantell and two other pilots were sent aloft in small planes to search for the strange object. Soon Mantell reported, "I've sighted the thing—it looks metallic and it's tremendous in size." The other two pilots failed to see anything and returned to base. Mantell continued to climb through broken clouds to an altitude of about 20,000 feet. Some hours later, search teams found his body and wrecked plane. It is now believed that Mantell was chasing a Navy Skyhawk balloon, a huge balloon used for high altitude research, then a Navy secret. It is likely that Mantell, whose plane had no oxygen equipment, flew too high, fainted, and crashed.

Inevitably, popular belief grew that flying saucers were invaders from outer space. It was theorized that some unexplained sightings of space vehicles could be intelligent extraterrestrial beings observing the earth and its inhabitants. A leading authority on UFOs, Philip J. Klass, punctures this fanciful notion by declaring, "My intensive studies have failed to uncover a shred of evidence that would prompt me even to suspect that the earth has had visitors from another world."

Controversy about UFOs led the United States Air Force to undertake an intensive investigation. The military was concerned because of the fear that flying saucers might be a new weapon of some foreign power. But after the Air Force had investigated UFOs for several years, it became clear that the entire effort was nonproductive. Nothing worthwhile had come out of the case and there was no indication of a threat to the national security, which was the Air Force's only concern.

In response to criticism about its lack of faith in UFOs, the Air Force contracted with the University of Colorado for a major new $500,000 scientific study of Unidentified Flying Objects, headed by a noted physicist, Edward V. Condon. It took more than thirty scientists and engineers to prepare the Condon Report, which analyzed 7,641 cases collected by the Air Force. The principal categories were classified as astronomical phenomena (including meteors, stars, and planets), aircraft, balloons, and satellites. A miscellaneous group of 1,031 cases included hoaxes, rockets, reflections, flares, fireworks, mirages, searchlights, clouds, and birds.

The Condon Report, published in 1969, concludes that "further extensive study of UFOs probably cannot be justified in the expectation that science will be advanced thereby." This judgment was supported by John Sladek, who comments in *The New Apochrypha*, "Under the presence of objective investigation, UFOs have been retreating, like unicorns, deeper into the forests of fantasy. There is no question but that the majority of UFO sightings are mistakes of one kind or another, built up by natural curiosity, a convenient myth, and the action of rumor."

The planet Venus has so often been mistaken for some sort of spacecraft, that she has been called the "Queen of the UFOs." One prominent person who was misled by sighting Venus was President Jimmy Carter. After a speech by Carter to a Lions Club meeting in Leary, Georgia, on January 6, 1969, he and others saw what they believed was an UFO, but was actually Venus, a conspicuous evening star nearing its maximum brilliance. Carter was in good company in misinterpreting Venus as an UFO. During World War II, many rounds of ammunition were wasted firing at Venus, in the belief that it was an enemy aircraft. Actually, Venus was thirty million miles away.

Carl Sagan, another eminent astronomer, writes that "most UFOs have been identified as belonging to one of the following categories: unconventional aircraft; aircraft under uncommon weather conditions; aircraft with unusual external light patterns; meteorological and other high-altitude balloons; artificial earth satellites; flocks of birds; reflections of searchlights or headlights off clouds; reflection of sunlight from shiny surfaces; luminescent organisms (including one case of a firefly lodged between two adjacent panes of glass in an airplane cockpit window); mirages; lenticular cloud formations; ball lightning; sun dogs; meteors, including green fireballs; planets, especially Venus; bright stars; and the aurora borealis." In short, most reported UFOs are identified as natural though unusual occurrences, wrongly interpreted.

Martin Garner's *Fads and Fallacies in the Name of Science* states that "there is every reason to expect that the saucer mania will go down in history as merely one more example of a mass delusion."

A fine example of such a mass delusion occurred on Halloween night in 1938, when Orson Welles presented a radio show based on H. G. Wells's *War of the Worlds*. The broadcast opened with dance music interrupted by a series of news flashes. A gas explosion on Mars was reported, followed by an earthquake tremor in New Jersey, and finally a broadcast from Grover's Mills, New Jersey, describing the landing of a huge, cylindrical spaceship out of which came bug-eyed monsters armed with death rays. Other cylinders landed, and the monsters destroyed New York before they died of disease germs for which they had no resistance. Mass panic followed from coast to coast. Thousands wept, closed their windows to shut out poison gas, or fled from their homes expecting the world to end. Hysteria was especially prevalent in the southern states among the poorly educated.

In the light of the Orson Welles hoax it is easy to understand the popular belief in flying saucers.

A new approach to the flying saucers discussion comes from psychologist Carl Gustav Jung. Jung's theory is that flying saucers are psychic projections or "visionary rumors" which reflect a deep-seated air of disquiet among humanity. With the old gods dying, if not dead, and the world menaced by threat of total destruction as never before in its history, men are turning to the skies in search of their redeemers.

A two-volume bibliography, *UFOs and the Extraterrestrial Contact Movement*, by George Eberhart, was published in 1986. The 15,600 entries include 5,500 books in English, 2,200 books in other languages, and several periodicals. One entry suggests that Adolf Hitler and other Nazis escaped from Germany in 1945 and have been surviving in a colony at the South Pole, which is their base for an air force of flying saucers. Another listing is a record entitled *Music from Another Planet*, songs taught by a man from Saturn. One listing is a chapter on UFO abductions, reports from people claimed to have been kidnapped by aliens and later released. Another is on animal mutilations where extraterrestrials are suspected of killing cattle and surgically removing organs. Chronologically, there are cases dating back 2,500 years to the Roman Empire.

The most thorough analysis and debunking of UFOs is Philip J. Klass's 1983 study entitled *UFOs: The Public Deceived.* The need for such a work is shown by the widespread popular belief that intelligent beings exist elsewhere and have visited the earth. A Gallup poll in mid-1978 revealed that 57 percent of adult Americans believe that UFOs are "real."

In reality, Klass finds that UFOs are a myth. The author has been a leading investigator of UFO reports and an authority on the subject for the past twenty years. He concludes that the vast majority of UFO reports are misidentified natural phenomena and manmade objects, misunderstandings of the limitations of radar-electronic equipment, misunderstandings of changing atmospheric conditions and optics, unreliability of eyewitness testimony, and overactive imaginations. UFO reports often turn out to be stars or planets, privately operated helicopters, or perhaps meteors or birds. Klass concludes "that roughly 98 percent of all UFO reports are simply misidentifications of prosaic, if sometimes unfamiliar objects by basically honest observers, and that the balance, roughly 2 percent, are self-delusions or hoaxes by persons who like to spin tall tales and become instant celebrities."

In reference to those who for various reasons wish to keep the UFO myth alive, Klass declares, "The greatest derangement of the mind is to believe in something because one wishes it to be so."

A new Gallup survey taken in 1986 found that 50 percent of the respondents expressed the belief that there are "people somewhat like ourselves living on other planets in the universe," while 34 percent were skeptical, and 16 percent were unsure. Nine percent reported they had actually seen something they thought was an UFO. Belief in the existence of life on other planets and in UFOs was substantially higher among people who had attended college than among those whose education ended in high school or below. Also, men were more likely to believe in extraterrestrial life than were women. Skepticism grew with age. People over fifty were less likely than young adults to be convinced of UFOs and extraterrestrial life as facts.

Perhaps the final word on the subject comes from Christopher Evans, in his *Cults of Unreason.* "Despite all the denials of Air Force generals and the tempered skepticism of specialist scientists, flying saucers and their superior occupants, whether sacred or profane, exist, have existed, and probably always will exist ... if UFOs did not exist it would have been necessary to invent them."

REFERENCES

Condon, Edward V. *Scientific Study of UFOs.* New York: Dutton, 1969.

Eberhart, George, and J. Allen Hynek. *UFOs and the Extraterrestrial Contact Movement.* New York: Garland, 1986.

Evans, Christopher. *Cults of Unreason.* New York: Farrar, 1973.

Hynek, J. Allen. *The UFO Experience.* Chicago: Regnery, 1972.

Klass, Philip J. *UFOs: The Public Deceived.* Buffalo, N.Y.: Prometheus Press, 1983.

Klass, Philip J. *UFOs Identified.* New York: Random House, 1968.

Menzel, Donald H., and Ernest H. Tares. *The UFO Enigma: The Definitive Explanation of the UFO Phenomenon.* Garden City, N.Y.: Doubleday, 1977.

Sagan, Carl. *UFOs: A Scientific Debate.* Ithaca, N.Y.: Cornell University Press, 1973.

Sheaffer, Robert. *The UFO Verdict.* Buffalo, N.Y.: Prometheus Books, 1981.

CHAPTER

27

Extrasensory Perception—ESP

The term *extrasensory perception* was invented by an American psychologist, Joseph Banks Rhine, in the early 1930s. Reports of ESP phenomena have been recorded since biblical times, but they were not subjected to serious scientific research until the founding of the Society for Psychical Research in London in 1882. Since then the Society's *Proceedings* and *Journal* and similar publications in the United States, Holland, and other countries have reported research findings.

Three types of ESP phenomena have been investigated. They include telepathy, that is, a direct response to another person's thoughts; clairvoyance, i.e., the direct response to an object or event; and retrocognition, a response to some past event or no-longer-existing object. Another area of study in the field known as parapsychology is psychokinesis, dealing with the movement of physical objects by the mind and survival after bodily death.

Early ESP research was mainly concerned with investigating the authenticity of reported cases and studying individuals who appeared to be psychic. The scientific community in general was highly skeptical when facing unseen phenomena such as extrasensory perception, telepathy, and clairvoyance. Those areas of parapsychology were regarded as so much hocus-pocus, belonging on a carnival stage rather than in a laboratory.

There was a slight softening of that attitude in 1969 when the Parapsychological Association was admitted to the American Association for the Advancement of Science, but recently there was an unsuccessful attempt made to oust the parapsychologists from the AAAS, and there is limited attention to the field in academic institutions.

Genuine ESP research really began with the arrival at Duke University of J. B. Rhine. Rhine was a native of Pennsylvania, served two years in the Marines, graduated from the University of Chicago, taught at West Virginia University, and received a doctorate in botany from the University of Chicago. In the early twenties Rhine and his wife, Louisa E., heard and were deeply impressed by a lecture on spiritualism given by Sir Arthur Conan Doyle. There followed a long period of study of the literature and practice of spiritualism by Rhine and his wife. In 1927, Rhine became a research assistant at Duke, working on psychic forces under William McDougall, formerly of Oxford and Harvard. Rhine joined the Duke faculty in 1928 and in 1940 became director of the school's Parapsychology Laboratory.

For his experiments, Rhine adopted a set of cards consisting of five symbols: a square, a circle, a cross (or plus sign), a five-pointed star, and a set of three wavy lines. The subject whose ESP ability was being examined was asked to guess or "call" the order in which the symbols appeared in the pack of twenty-five cards. The person in charge of the experiment successively placed each card face down on the table in front of him and recorded the subject's guess as to which symbol it bore. If a subject scored so well that his score could be expected to occur by chance only once in a thousand times, Rhine held that he was showing "extrachance" ESP.

A charge that Rhine's statistical procedures were unsound was denied by the president of the American Mathematical Association, who approved of the statistical methods used.

Critics attacked Rhine's experimental procedures on another ground. For example, the standard ESP cards were printed in such a way that if a bright light fell on one at a certain angle, the symbol could be read through the back of the card. Other possibilities might be a chance exposure of a card, a reflection of a card in the experimenter's glasses, or an unconscious reaction at the sight of certain symbols. To avoid such happenings, the subject and the experimenter were located in separate rooms.

There were also occasional charges of fraud, but these were not seriously considered if similar results were reported in two or more laboratories.

In later experiments, Rhine turned to dice in place of cards. Critics have pointed out the dice are subject to the same pitfalls of statistical error and unconscious selection that are involved in card testing.

An intriguing question was raised by Rhine in an article published in 1951. Do animals possess powers of extrasensory perception? He was convinced that "Lady," a professional mind-reading horse in Richmond, Virginia, had psychic powers. He cites also the case of a dog in California that howled all day and stopped just as an earthquake occurred at Long Beach, and a collie that whimpered under a bed until an explosion at a nearby plant killed his master. In both cases, Rhine thought the dogs had precognition of coming events. Many studies of animal behavior have centered on chimpanzees. Comparative psychologists insist that chimpanzees have the capacity for self-awareness, self-consciousness, and self-knowledge, but doubters maintain that everything chimpanzees do should be dismissed as mimicry, cuing by their human handlers, poorly designed experiments, or simple stimulus and response, in short, attributing human characteristics to animals.

Some investigators are convinced of the existence of psychic powers, but believe that these powers are difficult to prove and to understand. Gardner's *In the Name of Science* concludes that "The most damaging fact against Rhine is that, with very few exceptions, the only experimenters who have confirmed his findings are men who share his strong belief in psychic phenomena. Hundreds of tests by doubting psychologists have been made, and yielded negative results." Rhine's defense is to blame the attitude of the experimenters, maintaining that their negative beliefs have a marked influence on the subject.

Another analyst, C. E. M. Hansel, in his *ESP: A Scientific Evaluation*, summarizes the facts regarding ESP as follows:

1. Subjects, when attempting to guess card symbols, have attained scores that could be attributed to chance.

2. Some of those taking part in ESP experiments have indulged in trickery.

3. Subjects who attain high scores cannot do so on all occasions.

4. Subjects tend to lose their ability to obtain high scores.

5. A successful subject is sometimes unable to obtain high scores when tested by a critical investigator.

6. A subject may attain high scores under one set of experimental conditions and fail to do so under other experimental conditions.

Hansel concludes that "A great deal of time, effort, and money has been expended but an acceptable demonstration of the existence of extrasensory perception has not been given." On the other hand, many prominent philosophers, psychologists, and writers have accepted ESP, among them William James, Jules Romains, H. G. Wells, Maurice Maeterlinck, Arthur Conan Doyle, Aldous Huxley, Gilbert Murray, Upton Sinclair, and Arthur Koestler. Among scientists, Thomas Edison and Luther Burbank were believers in telepathy.

A recent report on the attitude of scientists toward ESP and parapsychology was published in the *Skeptical Inquirer* in 1983. Committee chairmen and council members of the American Association for the Advancement of Science were polled, a total of 497 AAAS leaders considered to be a scientific elite. Only 4 percent of the respondents thought ESP to be "an established fact." An additional 25 percent regarded it as a likely posibility. Thus only 29 percent of the scientists were favorably disposed toward ESP. Fifty percent considered ESP either "a remote possibility" (41 percent) or "an impossibility" (9 percent). The remaining 21 percent rated it "merely an unknown."

This survey is in striking contrast to a poll of readers conducted by the English weekly *New Scientist* ten years earlier. At that time there was a surprisingly high level of belief. Some 73 percent of the 1,216 respondents believed that ESP was either an established fact or a likely possibility. A Gallup poll in 1978 found that 51 percent of adults surveyed believed in ESP.

The elite scientists gave two explanations for their negative attitude toward ESP: first, there is insufficient evidence for psychic ability, and second, parapsychological research has not been conducted in a competent manner. It is clear that parapsychologists will have to produce more solid evidence and show greater scientific sophistication before their claims will be accepted by the scientific community.

REFERENCES

Cohen, Daniel. *ESP: The Search Beyond the Senses.* New York: Harcourt, 1973.

Ebon, Martin. *Test Your ESP.* New York: New American Library, 1971.

Gudas, Fabian. *Extrasensory Perception.* New York: Scribner, 1961.

Hansel, C. E. M. *ESP: A Scientific Explanation.* New York: Scribner, 1966.

Holer, Hans W. *The Truth about ESP.* Garden City, N.Y.: Doubleday, 1974.

Knight, David C. *The ESP Reader.* New York: Grosset & Dunlap, 1969.

Murchison, Carl. *The Case for and against Psychical Belief.* Worcester, Mass.: Clark University Press, 1921.

Pratt, Joseph G. *Extra-Sensory Perception after Sixty Years.* Boston: Bruce Humphries, 1966.

Rhine, Joseph B. *Extra-Sensory Perception.* Boston: Humphries, 1964.

Schmeidler, Gertrude B. *Extra-Sensory Perception.* New York: Atherton Press, 1969.

Sherman, Harold M. *How to Make ESP Work for You.* Los Angeles, Calif.: DeVores, 1966.

Smith, Susy. *ESP.* New York: Pyramid Books, 1962.

Walker, Kenneth R. *Extra-Sensory Mind.* New York: Emerson Books, 1965.

CHAPTER

28

Search for Medical Miracles

It is estimated that medical quackery in violation of U.S. laws costs Americans $2 billion a year—more than all other swindles and confidence games combined, and more than is spent on the scientific study of disease. Pseudoscientists have flourished in the field of medicine since time immemorial. They often make a great deal of money and advertise freely in newspapers, in magazines and on television. A distressing feature, sometimes fatal, is that their patients delay treatment by reputable physicians until it is too late.

A number of medical cults flourish in the United States and attract millions of devoted followers. The first such alternative form of treatment, homeopathy, originated in 1790 with a German doctor, Samuel Christian Hahnemann, who published a book entitled *Organon* in 1810. The basic idea was that "like cures like," i.e., a drug will cure a disease if the same drug, taken by a healthy person, will produce symptoms similar to those of the disease. The remedies were to be taken in extremely small doses, on the theory that the more minute the dose the more potent it would be.

Homeopathy was a strong reaction against the use of worthless drugs prevailing in Hahnemann's time. Botanical and other valueless concoctions were numerous. Homeopathy's simple remedies won immediate popularity. The founder was superstitious, for he believed that plants possessed spiritual powers which could be released if properly treated. These powers could be released by dilution. Futher, the therapeutic activity of any drug would decrease with its greater concentration. Only the symptoms of a disease could be successfully treated, and medical treatment must be supplemented by appropriate and hygienic measures.

Homeopathy reached its peak of popularity in the early nineteenth century, with schools, clinics, societies, and journals devoted to its promotion. By the middle of the twentieth century, however, the practice of homeopathy had almost disappeared. While still popular, homeopathic medicines numbered about 3,000. Some idea of their value may be estimated by naming a few: tears from a weeping young girl, powdered starfish, skunk secretion, crushed live bedbugs, powdered anthracite coal, powdered oyster shells, uric acid (obtained from human urine or snake excrement), and medicines made from various types of spider webs and poisons. All these queer drugs have been tested by professional pharmacologists and found harmless (and useless) if taken in minute diluted form. Today, there are no homeopathic schools surviving, though several thousand doctors still consider themselves homeopaths. All have standard M. D. degrees. Since World War II, there has been a revival of homeopathy in Europe, especially in Germany and France. According to Martin Gardner, science writer and author of *Fads and Fallacies in the Name of Science*, "The cult has always been popular among the nobility of England and the Continent. India and South America are other areas where the movement is flourishing."

One noted homeopath, Charles H. Duncan, published a book, *Autotherapy* (1918), in which he explained how he cured boils by giving the patient an extract from the boil, tuberculosis by an extract from the sputum, and dysentery by injecting a fluid obtained by filtering the excretions.

Another famous irregular branch of medicine, naturopathy, originated in Europe but spread world wide. Its basic claim is to rely on nature for healing, with little or no resort to medicine and surgery. Pioneers in the field recommended such diverse measures as water cures, sleeping on bare ground, walking barefoot on wet lawns and sand, steam baths, sunlight, a vegetarian diet, and whole wheat bread.

One of the great promoters of naturopathy, Bernard MacFadden, author of a monumental five-volume *Encyclopedia of Physical Culture*, listed many minor diseases that could be treated at home with special diets, exercises, and water therapy. MacFadden was convinced that cancer could be cured by a diet of nothing except grapes.

A branch of naturopathy is known as zone therapy. For this purpose, the body is divided vertically into ten zones, five on each side of the body, with each zone terminating in an individual finger and toe. All types of body pain could be relieved and usually removed by pressure on the proper finger or toe or some part of the affected zone.

The naturopath is adamantly opposed to the bacterial theory of disease—a rejection shared by many religious groups: Christian Science (founded by Mary Baker Eddy), New Thought, Unity, and Jewish Science. The most famous individual to oppose the germ theory was George Bernard Shaw, who in one of his last books, *Everybody's Political What's What*, defends naturopathy. In Shaw's opinion, most epidemics could be traced to laundries where microbes from one person's clothing infected others. Quite in character, Shaw also opposed vaccination, vivisection, eating meat, Caesarian operations, and the removal of tonsils and appendixes.

A favorite target of naturopaths is vaccination. The trials of Edward Jenner, who developed vaccination to control smallpox, are described in the introduction. The astounding health gains produced by vaccination in recent years are widely recognized. Smallpox is believed to be wiped out world wide. Whooping cough is almost unknown, and polio, diphtheria, measles, and typhoid fever have been dramatically reduced, with the possibility of complete elimination in another decade. Pasteurization of raw milk has been a major factor in eliminating epidemics of scarlet fever, tuberculosis, and typhoid.

A third leading medical subculture is osteopathy, invented in 1874 by what Gardner calls "a medical illiterate" named Andrew Taylor Still. The theory was that practically every ailment resulted from a blockage of an artery or nerve. The blockages could be traced directly to the spine. When spinal vertebrae became dislocated, they caused lesions or subluxations. The remedy was to massage the spine, thereby adjusting and getting rid of the lesions. This treatment, according to Still, could cure tuberculosis, epilepsy, gallstones, tumors, yellow fever, malaria, diphtheria, rickets, and diabetes. The existence of germs was never admitted.

Spinal adjustments are still the specialty of osteopaths. The number of practitioners of osteopathy today is estimated at 13,000. The field has expanded considerably since the founder's day. Currently, osteopaths massage other parts of the body, give water and electrical therapy, administer anesthetics, prescribe drugs, remove tonsils, and deliver babies. The recognition and treatment of skeletal malrelationships, strains, and poor postures continue to be important aspects of osteopathy.

The founder of the fourth major medical development in America—chiropractic—was Daniel D. Palmer. Chiropractic arose about twenty years after osteopathy. Like the latter, chiropractic taught that most human ills are caused by spinal maladjustments. Chiropractors believe that when people become sick or disabled, the cause is nerve roots being pressured against the backbone. The cure is to manipulate, twist, or massage the spine back into line.

Palmer started in 1894 by practicing magnetic healing, treating people by "animal magnetism," but then he apparently picked up the idea of subluxations from the osteopaths. Today, chiropractors are the most widely known and most numerous of medical cultists in the United States. Their national organization claims 25,000 members in active practice with four million patients annually. A typical chiropractic spinal chart has parts labeled to indicate the areas of the body they control. Gardner comments that "These charts have about the same relation to modern anatomy as a phrenological chart of head bumps bears to modern brain study." Like naturopaths, chiropractors reject drugs and the germ theory of disease.

An ancient technique of Oriental origin, acupuncture has been receiving considerable attention in recent years. Acupuncture has been used to treat disease in China since about 2500 B.C. Since the late 1950s, the method has been used as an anesthetic by inserting needles into certain parts of the body. After insertion, the needles are twirled by hand or stimulated electrically. They are made of fine steel, gold, or silver. In Chinese theory, the puncture in the skin allows escape of noxious substances. Acupuncturists claim to cure arthritis, rheumatism, menstrual disorders, migraines, asthma, hay fever, and facial paralysis. Research has shown the claims to be groundless.

Acupuncture has been viewed by western physicians with a mixture of hope and skepticism. Some have noted that the method has merit for analgesic and anesthetic purposes, and think it deserves further experimentation and research. Other medical experts maintain that acupuncture is little more than quackery.

A medical technique once in common vogue but now discredited is bloodletting, a bizarre custom dating back to ancient Rome. Physicians of the Alexandrian School regarded excess of blood (Plethora) as the chief cause of diseases such as coughing of blood, epilepsy, pneumonia, tonsillitis, etc. Bloodletting was habitually applied to almost every condition.

There is convincing evidence that George Washington's death was caused by bleeding by his doctors. William Cobbett, an English political writer then living in the United States, attacked Dr. Benjamin Rush, a famous physician and well-known Democratic politician, accusing him of killing Washington with his special bleeding treatment and killing his other patients with excessive bloodletting. Rush sued for libel. The trial in 1799 ended with Cobbett being fined $5,000, his printing business destroyed, and his assets forced into bankruptcy.

Similar to George Washington's tragic end was the treatment of the dying King Charles II of England. Physicians cut off his hair, bled him, applied plasters of pitch and pigeon dung to the bottoms of his feet, poured antimony and sulfate of zinc down his throat, and fed him "spirit of human skull"—whatever that was. It took medical science five days to kill the king.

In *The New Apocrypha*, John Sladek concludes that the success of pseudomedicine surpasses that of any other crank system. "Millions of people who chuckle over flat earth notions still wear copper bracelets to ward off arthritis, rub on worthless hair restorers, allow quacks to pummel their spines, swallow unnecessary tonics and vitamins, avoid meats or foods of a certain color, and visit faith healers."

The most recent plague to strike mankind world wide, reaching near epidemic proportions in Africa and the United States, is AIDS (acquired immune deficiency syndrome), an almost invariably fatal disease. Scientists are under tremendous pressure to find a cure or an effective vaccine to halt the progress of the devastating disease. Much of the money being spent on dubious cures might be better spent on AIDS research.

Modern physicians are ambivalent in their attitudes toward such alternative forms of treatment as homeopathy, osteopathy, chiropractic, and acupuncture. They generally consider, however, that despite numerous practitioners and believers, these developments are not in accord with modern medical discoveries and science.

REFERENCES

Ackernecht, Erwin H. *A Short History of Medicine.* New York: Ronald, 1955.

Atkinson, Donald T. *Magic, Myth and Medicine.* Cleveland, Ohio: World Publishing Co., 1956.

Cartwright, Frederick. *Disease and History.* New York: Crowell, 1922.

Dekruif, Paul H. *Men against Death.* New York: Harcourt, 1932.

Haggard, Howard W. *Devils, Drugs, and Doctors.* New York: Harper, 1937.

Haggard, Howard W. *Mystery, Magic, and Medicine.* Garden City, N.Y.: Doubleday, 1933.

Inglis, Brian A. *History of Medicine.* Cleveland, Ohio: World Publishing Co., 1965.

Singer, Charles J. *From Magic to Science.* New York: Boni, 1958.

Sladek, John. *The New Apocrypha; A Guide to Strange Science and Occult Beliefs.* New York: Stein and Day, 1973.

Thomen, August A. *Doctors Don't Believe It.* New York: Simon & Schuster, 1941.

Afterword

Historical and laboratory researches have found answers to some of the mysteries posed in *Scientific Enigmas.* Others may continue to be matters of speculation and conjecture.

It is evident from reviewing the variety of subjects dealt with in the present work that there is scarcely any limit to popular beliefs in pseudoscientific ideas. Wide public acceptance exists, for instance, for astrology, flying saucers, missing continents, the Bermuda Triangle, sea monsters, spiritualism, extrasensory perception, and miraculous medical cults. Furthermore, many people take pleasure in bizarre beliefs and are reluctant to recognize truth. Entire publishing industries revolve around exploiting paranormal fads and fancies.

Scientists, on the other hand, are discouraged by the limited sale of debunking books trying to correct popular errors and misconceptions. Regardless of facts, many people continue to be credulous on slight evidence. Public interest in psychic phenomena, the occult, and pseudoscience continues. A prominent scientist, Paul Kurtz, points out that discredited ideas tend to recur, giving astrology as an example. By 1900, astrology was viewed as a historical curiosity, accepted only by the uneducated. Then there was a revival of interest "and even supposedly sophisticated people claim to believe in it." Some UFO enthusiasts insist that government authorities have censored news about flying saucers that proves their reality.

For centuries, gloomy prophets have been forecasting the end of the world, often setting exact dates for Armageddon. Each time the fatal day has arrived, mankind has been spared. The noted historian of science Isaac Asimov points out that the death of the sun will occur in about ten billion years when its hydrogen is used up, thereby destroying the earth. Thus, there is no immediate cause for concern.

A more immediate prospect, as seen by prophets of doom, is that man will destroy the world by overpopulation. The world's population has expanded from two billion people fifty years ago to more than five billion today. By the year 2009, there will be between seven and eight billion people on earth, requiring that the world's food supply be doubled if a substantial proportion of mankind is not to face starvation. The obvious solutions are international measures to curb the birth rate and improved technology to increase the food supply.

Another acute problem confronting mankind is decreasing energy sources. Petroleum, gas, and coal resources are being rapidly depleted. The most hopeful alternative source appears to be solar energy, which if possible to exploit fully would be virtually inexhaustible. The vast resources of the oceans, which span 63 percent of the earth's surface, are just beginning to be exploited.

Still, there are many grounds for hope. Medical research has virtually eliminated such formerly devastating killers as smallpox, poliomyelitis, and typhoid fever, as well as a variety of childhood diseases once responsible for high infant mortality. The bubonic plague or Black Death, which wiped out a third of Europe's population in the fourteenth century, would be impossible in the twentieth century. It is unlikely, too, that an epidemic such as Spanish influenza could again wreak the havoc it caused during the World War I period, although the recent rapid spread of AIDS (acquired immune deficiency syndrome) is creating world wide concerns.

It remains a mysterious world, as Arthur C. Clarke found in his work by that title. There will be no shortage of new and different scientific and social enigmas waiting to be solved by citizens of the future.

GENERAL REFERENCES*

Clarke, Arthur C. *Mysterious World.* New York: A. & W. Publishers, 1980.

*The Greenhaven Press of St. Paul, Minnesota, has announced a new series of books entitled Opposing Viewpoints, dealing with some of the topics included in *Scientific Enigmas*: dinosaurs, pyramids, Bermuda Triangle, solar system, Bigfoot, Noah's Ark, Loch Ness Monster, Stonehenge, and other phenomena.

Evans, Christopher. *Cults of Unreason.* New York: Farrar, 1973.

Gardner, Martin. *Fads and Fallacies in the Name of Science.* New York: Dover, 1957.

Gardner, Martin. *In the Name of Science.* New York: Putnam, 1952.

Goran, Morris. *The Modern Myth.* New York: Barnes, 1978.

Hering, Daniel W. *Foibles and Fallacies of Science.* New York: Van Nostrand, 1924.

Jastrow, Joseph. *Error and Eccentricity in Human Belief.* New York: Dover, 1935.

Jones, William. *Credulities Past and Present.* London: Chatto, 1880.

McKay, Charles. *Extraordinary Popular Delusions and the Madness of Crowds.* New York: Farrar, 1932.

Morgan, Chris, and David Langford. *Facts and Fallacies.* New York: Wiley, 1981.

Reader's Digest. *Strange Stories, Amazing Facts.* Pleasantville, N.Y.: 1987.

Sladek, John. *The New Apocrypha: A Guide to Strange Science and Occult Beliefs.* New York: Stein and Day, 1973.

Snow, Edward. *Supernatural Mysteries.* New York: Dodd, 1974.

Thomsen, J. Arthur. *Riddles of Science.* New York: Liveright, 1932.

Welfare, Simon. *Arthur C. Clarke's Mysterious World.* New York: A. & W. Publishers, 1980.

Index